T0181805

Agricultural Biotechnology in China

Valerie J. Karplus · Xing Wang Deng

Agricultural Biotechnology in China

Origins and Prospects

Foreword by Norman E. Borlaug

 Springer

Valerie J. Karplus
Technology and Policy Program
Massachusetts Institute of Technology
77 Massachusetts Ave.
Building E40-371
Cambridge, MA 02139
vkarplus@aya.yale.edu

Xing Wang Deng
Department of Molecular, Cellular, and
Developmental Biology
Yale University, P.O. Box 208104
165 Prospect St., OML 352A
New Haven, CT 06520-8104
xingwang.deng@yale.edu

ISBN: 978-1-4939-5062-1   ISBN: 978-0-387-71139-3 (eBook)
DOI 10.1007/978-0-387-71139-3

Cover illustration: Rural landscape near Guanping Village in central China's Hunan province.

Printed on acid-free paper.

9 8 7 6 5 4 3 2 1

springer.com

For Guanping Village

Note on Chinese Names

Throughout the text, we spell Chinese names using the pinyin romanization system instead of the original Chinese characters. Using the mainland Chinese convention, we write the last name followed by first name, as in Deng Xiaoping. We make exceptions for scientists that are widely known outside of China by their Chinese or American first name followed by last name, for example, Dr. Ray Wu. In the Works Cited, sources are listed according to the format used in the original citation, usually by the author's last name.

Scope of Work

This book is limited in scope to developments that occurred in mainland China, and mostly those occurring from 1949 to the present. Throughout the text, we use the name "China" to refer to the mainland only.

Foreword

China's strong economic development of recent decades is built upon its successes in agriculture research and development over the past 50 years. From the mid-1950s to the early 1970s, internal campaigns and upheaval dwarfed the achievements of many dedicated Chinese scientists. However, by the late 1970s the results of modern agricultural research really began to bear fruit. For the next 20 years, agricultural policies encouraged greater initiative, innovation, investment, efficiency, and risk-taking—all to the benefit of Chinese agriculture. Rapid expansion in the use of high-yielding modern varieties and chemical fertilizers, improved water management, and more effective weed, disease, and insect control practices all occurred during this period.

Since the mid-1990s, however, China's agricultural progress has slowed. Rural wages are now lagging significantly behind those in the cities. Rural-to-urban migration is accelerating, and will likely involve over 200 million people by 2020. Managing this demographic change and keeping the agricultural economy and farmer incomes expanding are huge social and technological challenges that stand to affect Chinese and world history.

China's population is projected to grow to 1.6 billion by 2050. A larger and wealthier population is likely to double food, feed, and fiber demand by mid-century, while the arable land area is likely to fall by 20 percent, if not more because of urbanization, land degradation and water shortages.

New agricultural science and technology—with greater national coordination of provincial and regional research organizations—will be critical to meeting these challenges. Far-reaching policy changes will also be needed in agricultural tax policy, land tenure, and farmer education.

Karplus and Deng provide an excellent account of how recent developments in agricultural biotechnology in China may be the next big step in a long tradition of agricultural advances. I commend them for this outstanding piece of scholarship.

Dr. Norman E. Borlaug
Nobel Peace Prize Laureate

Preface

The path to Guanping, a remote mountain village in central China's Hunan province, invites its travelers back in time. Over hills and valleys the elusive dirt track winds almost beyond the reach of China's whirlwind economic development, arriving at clusters of wooden cottages that house around one hundred people. The surrounding terraced hillsides, reminiscent of images in ancient brush paintings, overflow with the year's rice crop. One of the co-authors of this book, Xing Wang Deng, was born in this village, and as a boy followed the same path for miles to attend the closest elementary and middle schools. We followed well-worn footprints on a visit to China's rice farming regions in the summer of 2005, tracing a path that has led ultimately to the authorship of this book.

A closer look around Guanping revealed that even in its relative isolation, life in this typical rice planting village was changing rapidly. The figures that emerged to greet us were either the very old or the very young. Most of the young adult population had left for higher paying jobs in towns and cities, joining China's growing migrant population. Rice paddies were sown with elite high-yielding hybrid varieties developed by leading breeders in the provincial capital. Grafted onto an otherwise isolated rural landscape were satellite dishes, a small dam nearby to supply electricity seasonally, fashion magazines, and bottles of carbonated soda, all signs of increasing linkages to the outside world.

This brief glimpse of rural China masks deeper challenges facing the country's farmers. Since its earliest days, agriculture has been inherently demanding on the environment. Only a modest fraction of China's land is arable, and increasing land productivity has been part of the quest for survival over thousands of years. Today, as China's population surpasses 1.3 billion and diets change to reflect growing prosperity, this quest has never been more important. Especially over the last century, efforts to increase agricultural yields have taken a severe toll on the environment. Rising applications of pesticides, fertilizer, and irrigation have left soils polluted, salty, and depleted of nutrients, while straining limited water and energy resources. The creeks near Guanping village are almost silent and lifeless, lacking the fish and shrimp that were once abundant. High value fertile land is being lost to nonagricultural uses, such as commercial development. These trends cast doubt on China's ability to provide an adequate, safe, and environmentally sustainable food supply to future generations.

A typical rural household in Guanping Village. Photo by authors.

As part of efforts to address these challenges, China's leaders are redoubling investment in science and technology programs. This emphasis on science and technology in China has deep historical roots. For thousands of years, leaders and farmers have relied on incremental advances in science and technology to overcome challenges facing the agricultural sector. The introduction of high-yielding crop varieties and associated inputs in the 1960s and 1970s increased land productivity and helped China to maintain food self-sufficiency through the early 1990s. In part due to the growing population and mounting environmental toll, the search is now on for new solutions. Agricultural biotechnology is one option that has emerged at the crossroads of the country's high technology and agricultural development agendas. We wrote this book to explore how support for agricultural biotechnology has become a dominant theme in China's agricultural research efforts, and how its development might affect the future of the country's agriculture and food supply.

China's rapid advances in the development of transgenic crops, which contain yield-enhancing or other useful traits transferred in by a set of molecular techniques, have drawn the most widespread attention. Several years ago, in the wake of food safety scares and widespread uncertainty, transgenic technology was widely shunned, particularly in parts of Europe and industrialized East Asia. Few observers realized that more than five years before the conflict erupted, virus-resistant transgenic tobacco was being planted commercially in China. Although the transgenic tobacco has since been withdrawn from the market, research on a wide array of other crops is now underway. Meanwhile, the widespread application of non-transgenic techniques aimed at solving some of the country's most persistent agricultural challenges has largely escaped public notice. While some stories of agricultural biotechnology in China have dominated headlines and others have not, we wrote this book to provide a more comprehensive picture of recent developments for citizens and policymakers, both in China and around the world.

We hope this book will find an eager audience in university students beginning their careers in an age of rapid transformation, as well as policy practitioners and the curious public in China and abroad. The number of opportunities available for biologists and biotechnologists in China is growing at an unprecedented pace. These individuals will shape the future of the field in China and its international reputation. China's agricultural biotechnology programs have been crafted with an eye to establishing the country as a major contributor in international scientific circles, boosting economic competitiveness in world markets, and modernizing its agricultural sector. Achieving these goals will depend on whether or not the research enterprise and commercialization channels are effectively managed. This book explores these management challenges, and offers several guiding insights for the future of agricultural biotechnology in China.

The completion of this book owes much to the efforts of our reviewers, colleagues, family, and friends. First, we would like to thank Carl E. Pray, Huang Jikun, Scott Rozelle and others affiliated with the Chinese Center for Agricultural Policy in Beijing. As leading researchers on the economic aspects of agricultural biotechnology in China, they have provided us with anecdotes and access to the center's data and manuscripts, which have proven invaluable in the completion of this work. We also extend our gratitude to the directors, staff, and students of the National Institute of Biological Sciences, Beijing for their support for this project from 2005 to 2006. We also thank Zhuang Qiaosheng, Fan Yunliu, Jia Shirong, Lin Min, and Guo Sandui at the Chinese Academy of Agricultural Sciences, and Zhang Qifa at the Huazhong Agriculture University, for tours, interviews, and reports of their work. We thank Chen Zhangliang and his colleagues at China Agricultural University for personal stories, connections to scientists, and photographs that have proven essential to understanding the early years of agricultural biotechnology in China. Thanks also to Li Xiaoyun, Xiang Ying, and the College of Humanities and Development at China Agricultural University for their help in organizing field studies in China's Henan and Hebei provinces in 2002. We are further grateful to Dr. Ray Wu at Cornell University and representatives of the China National Center for Biotechnology Development for their insights into early government support for China's biotechnology programs. We also appreciate the input of industry representatives and the staff at the Agricultural Affairs Office at the United States Embassy in Beijing. We thank Gary Toenniessen and Deborah Delmer at the Rockefeller Foundation for sharing the history of the foundation's support for biotechnology, particularly its applications to rice. Graduate students Chen Haodong and Liu Jianing rendered invaluable help with translation and research. Valerie extends a special thanks to the Henry R. Luce Foundation for supporting her work in China from 2002 to 2003.

Finally, we would like to thank those who assisted in the review, editing and preparation of this manuscript. We thank Huang Jikun, Rob Paarlberg, Carl E.

Pray, and Gary Toenniessen for valuable feedback on its concept and content. Our gratitude also belongs to Marianne S. Karplus, Susan S. Karplus, David L. Rager, and William R. Sherman for their encouragement and willingness to edit drafts in the final stages. Valerie extends heartfelt thanks to Paul L. H. Cook, Moira Heiges, Paul T. Karplus, Shan Liu, Nicolas Osouf, Jean D. Sherman, Jennifer Shuford, and David G. Victor for their support throughout the writing process.

We sincerely hope that you will enjoy reading a tale of how old and new have shaped the emergence of agricultural biotechnology in China. Though we cannot predict its outcome, we hope the following pages will illuminate the way forward.

Valerie J. Karplus and Xing Wang Deng

Contents

List of Figures

Note: Unless otherwise stated, photographs used in the cover and figures were taken by the authors.

List of Acronyms

ASQIQ – Administration for Quality Supervision, Inspection, and Quarantine (China)
Bt – *Bacillus thuringiensis*
CAAS – Chinese Academy of Agricultural Sciences
CAS – Chinese Academy of Sciences
CCAP – Center for Chinese Agricultural Policy
CIMMYT – International Maize and Wheat Improvement Center
CNCBD – China National Center for Biotechnology Development
CpTI – Cowpea trypsin inhibitor
CRTDC – China Rural Technology Development Center
DNA – Deoxyribonucleic acid
DPL – Delta and Pine Land Company
EPA – Environmental Protection Agency (United States)
EU – European Union
FDA – Food and Drug Administration (United States)
GDP – Gross Domestic Product
IARC – International Agricultural Research Center
IRRI – International Rice Research Institute
MOA – Ministry of Agriculture (China)
MOFTEC – Ministry of Foreign Trade and Economic Cooperation (China)
MOST – Ministry of Science and Technology (China)
NARC – National Agricultural Research Center
NDRC – National Development and Reform Commission (China)
NIH – National Institutes of Health (United States)
NNSFC – National Natural Science Foundation of China
SDPC – State Development Planning Commission (China)
SEPA – State Environmental Protection Administration (China)
SPC – State Planning Commission (China)
USDA – United States Department of Agriculture (United States)
WTO – World Trade Organization

Introduction

At the beginning of the twenty-first century, China's agricultural sector is facing daunting challenges. Population growth is slowing but will add nearly 300 million additional mouths before 2050, bringing the country's total population to around 1.6 billion. As deeper integration into the global economy requires China's farmers to compete directly with producers abroad, farmers are facing pressure to reduce costs and raise yields. Toxic pesticides harm tens of thousands of farmers each year, and residues persist in some final food products. Pesticide sprayings, together with increases in fertilizer applications, are further straining limited and fragile arable land.

Mindful that agricultural productivity has long influenced the country's growth and stability, China's leaders have searched for solutions to these contemporary challenges. During the late 1970s and 1980s, researchers around the world began to apply breakthroughs in the molecular-level understanding of plant function to develop crops with higher yield potential, better quality, and a reduced environmental footprint. These developments laid the foundation for modern agricultural biotechnology. In the 1980s, as laboratory advances abroad seemed to foretell economic and environmental benefits, agricultural biotechnology gained the support of many Chinese scientists and leaders. Eager to reap these benefits, China's leaders announced new state-funded programs that would provide an extensive capacity for agricultural biotechnology research.

This set of developments marked the beginning of agricultural biotechnology in China. Since the early 1980s, government funds have been used to send scientists abroad for training and establish positions at newly constructed modern laboratories to encourage their return. An intensive research effort has focused on developing crops suited for Chinese farms and domestic markets. Many laboratories employ the latest transgenic techniques (also known as genetic modification or genetic engineering), which involve the selective transfer of genetic material from one organism to another. Transgenic insect-resistant cotton varieties were developed and are now widely planted on a commercial scale. A Chinese team collaborated in sequencing the genome of one prominent rice subspecies and another team sequenced the genome of another subspecies independently. The number of scientific publications in top international journals published annually by mainland Chinese plant scientists between 1980 and 2006 steadily increased, most dramatically in last few years (Chen, Karplus, Ma, & Deng, 2006).

In the mid-1990s, the commercialization of the world's first transgenic crops sparked public opposition in many parts of the world. By the end of the 1990s, the outcry had reached a fever pitch. Much of the criticism centered on fears that control of the seed industry would be concentrated in the hands of a few large multinational companies. Many were concerned that farmers would no longer be able to save seed containing valuable patented material and would fall deeper in debt in order to be able to afford more expensive improved seed. Consumers in many developed countries asked why they should risk eating transgenic crops in the absence of direct benefits and proof that the crops were one-hundred percent safe. Transgenic techniques were occasionally confused with other (non-transgenic) approaches to crop improvement, dampening support for agricultural biotechnology research in general.

In the years following 1999, this backlash brought the commercialization of agricultural biotechnology products in Europe and Japan to a standstill. The controversy coincided with rising concern in much of the developing world about potential safety risks and economic dependency that might accompany the introduction of transgenic crops. In China, domestic approvals of transgenic food crops took a cautionary turn, in step with reactions abroad. A variety of insect-resistant transgenic rice passed all required health and safety tests, but as of mid-2007 had not received final commercialization approval. Aside from the recently reported approval of a virus-resistant transgenic papaya, no transgenic food crops have been approved for commercial planting in China in the last ten years. However, China allows imports of transgenic crops for food use, provided safety requirements are met.

Despite the slowdown in commercialization, China's leaders have nevertheless moved ahead with broad-based support for agricultural biotechnology research. Some laboratories are employing new insights into plant gene function to improve on traditional crop breeding methods. Other laboratories concentrate on developing transgenic crops, including varieties resistant to diseases and insect pests, or tolerant to drought or high salt conditions. Work in China's labs exceeds the scope of existing research on transgenic varieties in most developed countries. Most of the research in China is publicly funded and intended to address environmental and economic challenges facing China's agricultural system. Although this book focuses primarily on applications in plant crops (and focuses specifically in later chapters on transgenic crops), research efforts also include livestock and microorganisms. The growth of agricultural biotechnology in China has been paralleled by advances in biomedicine and driven by fundamental research in biochemistry, molecular biology, and genetics.

As of 2007, one transgenic crop, insect-resistant Bt cotton, was widely planted in China. Since 1997, over 100 varieties of Bt cotton have been released in China's cotton growing provinces, which occupy areas as geographically diverse as the northeast and the far west. Acreage planted to the new varieties began to expand once farmers realized yield gains and reduced pesticide inputs in the first years of adoption (Pray, Ma, Huang, & Qiao, 2001b). The introduction of Bt cotton has also raised several concerns. Some concerns are scientific—farmers have worried that the new transgenic cotton will lose its effectiveness against pests, just as pests developed resistance to pesticides sprayed on the non-transgenic cotton it replaced. Preserving the economic benefits to both producers and users over the long term is also a major concern. Weak intellectual property protection initially led to the widespread unregulated breeding and diffusion of the Bt trait, diluting its effectiveness. Still other concerns focus on safety. The government has developed regulations that establish procedures for initial safety approval and long term monitoring. However, enforcement of these provisions in China's large and decentralized agricultural sector has been uncertain, drawing criticism both domestically and abroad.

The past, present, and future of agricultural biotechnology in China is the subject of this book. Transgenic crop plants are part of a larger set of advances in the field of agricultural biotechnology, which has brought biotechnologists and ordinary farmers together in an unlikely union. This union looks quite different in China than in other parts of the world. A vast proportion of China's population consists of smallholder farmers whose livelihoods are directly impacted by agricultural technology. By contrast, in most advanced industrialized countries, farmers represent a small fraction of the population. Whereas transgenic crop plants in the United States and Europe mostly originated in the labs of private companies, public research institutes and universities are the primary developers in China. Relationships among stakeholders in China, including the government, research institutions, seed companies, and consumers, differ from their counterparts elsewhere in the world in ways that have affected the technology's development and adoption.

Several distinct features of the emergence of agricultural biotechnology in China make its story worth a closer look. First, agricultural biotech research and development in China have taken place predominantly in the public sector, whereas the private sector has taken the lead in other parts of the world. Will this affect whether or not the technology is accepted by farmers and consumers? Does a strong public sector role mean that development of new varieties will be primarily driven by public concerns, such as impact on the environment and rural welfare? Second, most of the research has focused on the prominent threats to agricultural yields in China, such as pest invasions and exposure to stresses, including drought or saline soils. Will more transgenic food crops ultimately be approved for planting in China, and if so, what impact will they have on the lives of China's millions of farmers and the productivity of the agricultural sector? Could agricultural biotechnology, as its proponents claim, hasten the transition to a more efficient and environmentally sound farming system? Or is it likely to founder amid a lack of incentives for commercialization, poor local safety monitoring, and stalled regulatory approvals?

Finally, despite only a few commercialization approvals, China's sizeable investment in agricultural biotechnology has produced significant results. These successes suggest an eastward shift in the frontier of innovation in the plant sciences and agricultural biotechnology. How will this shift affect China's competitiveness in world agricultural markets? What impact might it have on China's ability to influence policy related to the development of agricultural biotechnology, particularly transgenic crops, in other parts of the world?

These are questions of great importance and interest to scientists, students, economists, investors, and consumers around the world. If China's scientific research enterprise continues to grow in step with its economy, local advances in agricultural biotechnology will be increasingly felt in markets across the globe. Regardless of personal views on agricultural biotechnology, there is no disputing the extent of its development in China. This book offers policymakers and citizens a comprehensive picture of the origins of agricultural biotechnology in China, so that they might responsibly shape its future.

1

From Seeds to Empires: China's Long Agricultural History

China's agricultural sector has evolved over more than ten thousand years into one of the largest and most productive on earth. Until the mid-1900s, an overwhelming majority of the population worked in agriculture. Even today, the agricultural sector accounts for around 40 percent of the country's employment (*China Statistical Yearbook*, 2006). Although China has only seven percent of the world's arable land, its agricultural sector now provides over one fifth of humanity with a diet more abundant and varied than at any point in the country's history.

The stability and adequacy of China's food supply are relatively recent developments. For thousands of years, China's farmers labored long hours for uncertain harvests as they attempted to hold pest invasions, drought, floods, and soil degradation at bay. Over time, improvements in technology and infrastructure gradually helped to safeguard yields. These gains were offset by tax burdens that stripped farmers of all but what was required to ensure political loyalty, and miscalculations were often fatal for dynasties. Striking a balance between farm needs and state agendas proved critical to the survival of the ruling elite.

Today, trade, technology, and economic development are rapidly transforming age-old patterns of rural life. Yet the practices and relationships that developed and persisted over thousands of years still influence the path of this transformation. This chapter tells the story of how geography, ingenuity, and human need shaped China's agricultural development prior to the modern scientific age. A brief overview of historical context reveals the roots of many of the agricultural sector's past and present challenges, as well as the responses to these challenges, many of which are still relevant today.

Geography

Agricultural expansion in China has encountered diverse and often formidable geography. Starting from the capital in Beijing, the country's present border follows bustling eastern coastal urban centers south to the mountainous paddy-farming regions along the border with Southeast Asia. It skirts the towering peaks of the Himalayas before turning north and arcing back across the flat western and central

V. J. Karplus and X. W. Deng, *Agricultural Biotechnology in China.*
© Springer 2008

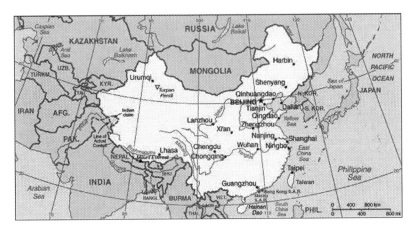

Fig. 1.1 Map of China. Reprinted from the *CIA World Factbook*, 2006.

steppe to the richly forested northeast, enclosing a total area of 9.6 million square kilometers, comparable in size to the United States (see Figure 1.1). China's two longest inland waterways, the Yangtze and Yellow Rivers, have long served as twin lifelines for agriculture in China's heartland.

The Yangtze River flows across China's midsection, tracing a temperate belt through the rice-growing regions of the western and central south, and nourishing fertile agricultural lands as far as the eastern coast. Several hundred miles north of the Yangtze, wheat overtakes rice as the dominant crop and covers large areas up to China's northeastern border. The Yellow River (*Huang He* in Chinese) begins at the northeastern edge of the Tibetan Plateau, and winds through one cradle of Chinese civilization on the North Central Plain, before emptying into the Yellow Sea.

Farming has long remained a challenge for China's agricultural communities. Much of the land area inside China's present borders is considered too mountainous or hilly to be suitable for cultivation. Both the Yangtze and Yellow Rivers for centuries proved a blessing and a curse to farmers living along their banks. Although the rivers provided abundant water and enriched soils, seasonal flooding could be extremely severe and often devastated villages and harvests. Especially in the north, limited rainfall and cold temperatures in winter left soils dry and cracked, and winds stripped away nutrient-rich topsoil (Myers, 1970). A lack of diverse soil types also limited farmers' ability to introduce new varieties that grew well in other places. On the southeastern coast, seasonal typhoons often claimed a significant portion of crop yields each year.

Earliest Roots: The Origins of Agriculture in China

Archaeological evidence suggests that China was home to at least one of the world's earliest agrarian civilizations. After 10,000 B.C., as nomadic groups were establishing the first agricultural settlements in the Middle East and Africa, China's earliest cultivated crops are thought to have emerged in several independent locations (Chang, 1999). Situated along the northeastern section of the Yellow River, the North China Plain emerged as an important center of origin for domesticated crops, including soybeans and several types of millet, a small-seeded species of cereal (Chang, 1986). Early settlers are also thought to have been among the first to cultivate the jujube, peach, apricot, persimmon, chestnut, hazelnut, and Chinese cabbage, as well as a wide variety of crops and trees for non-food uses (Li, 1983). Climate studies also suggest that temperatures and rainfall were higher than present levels, favoring cultivation (Chang, 1986).

Artifacts dating to around 6000 B.C. discovered along the Yellow River suggest that the Cishan, Peiligang, and related cultures established many of the earliest northern agricultural settlements. Peiligang burial sites revealed diverse remnants of probable food sources, which included grains, fruits, walnut, as well as the bones of pigs and dogs (Chang, 1986). Around the same time, some of the first known rice varieties were planted in China's tropical south. At the Pengtoushan site in Hunan province and the Hemudu site in Zhejiang province, among others, archaeologists discovered rice grains dating to earlier than 5000 B.C. that showed signs of domestication (Chang, 1999). Later cultures such as the Yangshao culture (5000 to 3000 B.C.) on the North China Plain engaged in a much broader array of cultivation activities, which included raising silkworms, planting hemp, and domesticating livestock (Chang, 1986). Thought to be descendents of the Peiligang, the Yangshao culture developed an array of primitive stone farm tools, techniques for pottery making, and methods for food storage (Chang, 1986).

Though traces of agricultural societies prior to 4000 B.C. abound, Chinese legend credits the mythical *Shen Nong* (or "Divine Farmer") with teaching the ancient Chinese the practices of agriculture in the third millennium B.C. (Needham, 1984). His instruction in agricultural practices and the classification of medicinal plants are described in several early Han Dynasty texts (Chang, 1999).

Rural Life in Early Imperial China: Xia, Shang, and Western Zhou Dynasties

Beginning with the first settlements on the North China Plain, rural life revolved around agriculture. Local cultures and lifestyles conformed to the rhythm of planting seasons. By closely observing weather patterns, day length, and the flowering of certain plants, early farmers along the Yellow River developed rudimentary forms of a calendar, which helped them to plan for optimal harvests. This system was increasingly replaced with calendars based on lunar cycles or solar terms (Needham, 1984).

As primitive farm tools, cultivation methods, and water control measures spread, agricultural yields increased. The Great Yu, the legendary founder of China's earliest dynasty (the Xia), is credited as the first to "tame" the Yellow River around 2200 B.C. by improving drainage (Chang, 1999). As farmlands expanded and came under the control of dynasties starting in 2000 B.C., protecting peasant households against economic and physical ruin became crucial to the leadership's survival. At junctures where several states vied for power, superior agricultural productivity could provide a critical trump card for an aspiring state.

As harvests grew more reliable, the inhabitants of China's earliest farming communities began to diversify their activities. Early forms of markets, written language, and government began to emerge once survival no longer required all hands in the field. Farmers began to exchange their harvests for goods and services in local markets. By the Shang Dynasty (1700 to 1027 B.C.), which succeeded the Xia around 1700 B.C., the earliest known Chinese writing system had emerged (Ho, 1975).

Though city work increasingly drew laborers away from ancestral lands, a strong link between the ruler and farmer remained. The elite class held power only as long as agricultural production could support it. Technological advances in agriculture helped some groups rise to power while others fell out of favor. Farmers depended on the government to enact flood control measures, as well as to establish and regulate agricultural markets. The state, in turn, owed its survival to agricultural tax revenues and its ability to shelter farmers from natural disasters and financial ruin.

As Chinese civilization expanded, the agricultural system was organized into units that could be governed and taxed by a central administration. During the Zhou Dynasty (1027 to 221 B.C.), the "well-field system" emerged, in which farmland was divided into units of nine plots, one of which belonged to the local nobility (Hsu, 1999). Eight families were assigned to farm the remaining plots, and shared responsibility for farming the noble's land. As the feudal system of the Zhou dynasty deteriorated, the previously subordinate Lu state was among the first to introduce land taxes. Officials collected taxes at the official rate of one-tenth of the crop, but at times the rate may have reached 20 percent (Hsu, 1999). This emerging system formed the basis for the small farm ownership and taxation that characterized China's agriculture for most of recorded history (Lewis, 1999).

Gains in agricultural productivity further enabled military campaigns and state expansion. Armies could reach only as far as available food stocks allowed, so agriculture was quickly expanded to conquered lands. To increase the total area under cultivation, states offered farmers incentives, such as tax reductions, irrigation infrastructure, and military protection. Expansion, by peaceful or coercive means, helped to relieve population pressure and avoid the endless subdividing of lands within families. Taxes on additional cultivated lands helped to boost state revenues, which strengthened individual states and provided impetus for further expansion (Lewis, 1999).

Benefits of Water Control, the Iron Plough, and Fertilizer: Eastern Zhou, Qin, and Han Dynasties

Despite early gains in agricultural productivity, pressures to increase food supplies persisted. The growth of urban centers increased the burden on farms to produce grain, not just for farm household consumption but increasingly for sale on emerging

Year	Period or Dynasty	Technology
10,000–2000 B.C.	Neolithic Period	Earliest records of cultivated rice farming.
2000–1500 B.C.	Xia	Primitive irrigation methods introduced.
1700–1027 B.C.	Shang	Earliest Chinese writing system developed.
1027–771 B.C.	Western Zhou	"Well-field system" emerged. Earliest records of taxation.
770–221 B.C.	Eastern Zhou 770–476 B.C. Spring and Autumn Period 475–221 B.C. Warring States Period	Bronze and iron ploughshares replaced stone models. First recorded use of natural fertilizer. Earliest known irrigation works in Hebei. *Dujiangyan* irrigation works built in Sichuan. Construction of dikes and levies along the Yellow River begins.
221–207 B.C.	Qin	First Qin emperor built a canal linking Yangtze and Pearl rivers, hastening state's rise.
206 B.C. to A.D. 9	Western Han	Iron farm tools grew in popularity. Irrigated area expanded.
A.D. 25–220	Eastern Han	Iron farm tools advanced and distributed with state support. Improved cultivation methods introduced.
A.D. 220–316	Western and Eastern Jin	
A.D. 420–588	Southern and Northern Dynasties	Agricultural text *Important Arts for the People's Welfare* published.
A.D. 581–617	Sui	Grand Canal construction completed.
A.D. 618–907	Tang	Quality and diversity of crops increased.
A.D. 960–1279	Northern and Southern Song	Early ripening rice introduced.
A.D. 1279–1368	Yuan	Cotton planting and weaving technologies improved.
A.D. 1368–1644	Ming	New World crops introduced. Major agricultural texts published.
A.D. 1644–1911	Qing	Advances in crop rotation and other cultivation techniques.
A.D. 1911–1949	Republic of China (mainland)	Increasing exposure to science and technology from Western countries.
A.D. 1949–present	People's Republic of China	Collectivization (1950s) and food self-sufficiency policies implemented.

Fig. 1.2 Milestones in China's Agricultural History.

markets. However, harvests were limited by nutrient-poor soils, especially in the north, as well as floods, soil erosion, desertification, pests, and diseases. A lack of transportation infrastructure also limited the redistribution of yields to meet the food demands of cities. In response, top-down and bottom-up efforts emerged to address these limitations to agricultural growth. Government officials again launched infrastructure projects and promoted technology they hoped would safeguard the food supply. Meanwhile, farmers also developed their own methods of water control, tillage, and seed selection to increase productivity.

Around 430 B.C., the first known large-scale irrigation project was built on the North China Plain (near present-day Hebei province) to channel water from the Yellow River to nearby fields (Lewis, 1999). Though it nourished crops along its banks, the Yellow River was prone to devastating floods, earning it the nickname "China's Sorrow" or "the Ungovernable." At various times throughout history, the deposition of silt has caused the bottom of the channel to be raised more than ten meters above the surrounding landscape, significantly higher than the earliest recorded levels, leaving the area especially prone to flooding (Giordano et al., 2004). Irrigation and water control measures only partially alleviated challenges to farm productivity. These measures were prone to failure; within a few hundred years shifts in the river's path and its heavy silt loads often overwhelmed them.

Under the Eastern Zhou Dynasty (which succeeded the Western Zhou and was named for the eastward position of its capital relative to its predecessor), several technologies dramatically increased the efficiency of farm labor. Perhaps most significantly, iron replaced stone as the primary material used in ploughshares, the sturdy leading edge of a plow (see Figure 1.3). Iron ploughshares enabled farmers to reduce plowing time and carve deeper furrows.

Under the latter half of Eastern Zhou rule, known as the Warring States period, several major waterworks projects were constructed. Developed by Li Bing, an official of the Qin state, and completed in 256 B.C., the *Dujiangyan* irrigation project effectively prevented flooding on the Min River near Chengdu city, opening a vast and fertile area for cultivation (see Figure 1.4). Comprised of a complex system of dikes and manmade channels, the project still irrigates approximately 500,000 hectares of land today (Song, 2001).

Throughout the Warring States period, usage of natural fertilizers spread. Various forms included green manures (in which live, often young, plants with high nutrient content were ground up and added to the soil). A variety of other substances not typically used in Western countries were also applied in China, including lime, mollusk shells, and soy cake (Needham, 1984).

Gains in agricultural productivity helped the Qin state to emerge victorious at the end of the Warring States period. Construction of the *Chengkuo* canal in 246 B.C. provided irrigation to 200,000 acres in central Sha'anxi province, and resulted in increases in production and population growth around the site of the future capital. Some scholars have suggested that irrigation played an important role in enabling the state's rise (Needham, 1984). The Qin Dynasty is credited with unifying China and establishing its capital at *Chang'an* (near present day

Fig. 1.3 Model plough with iron ploughshare outside the China Agriculture Museum in Beijing. The character engraved on the stone translates as "plough." Photo by authors. (A color version of this figure appears between pages 72 and 73.)

Fig. 1.4 Model of a water control structure used in the *Dujiangyan* irrigation system, now part of a United Nations World Heritage Site. Photo by authors. (A color version of this figure appears between pages 72 and 73.)

Xi'an) under the Emperor *Qin Shi Huangdi*. The Qin Dynasty was succeeded within less than a century by the Western Han, which popularized the use of iron tools and ploughs and expanded existing irrigation projects. Large-scale, centralized irrigation projects were particularly popular in the north, while in the south, citizens constructed tanks or reservoirs for individual household or village use (Needham, 1984).

Dynasties periodically sponsored massive infrastructure campaigns that improved transportation of agricultural goods, primarily between fertile lands in the south and major cities in the north. Perhaps the most prominent example was the construction of the Grand Canal. Begun in the fifth century B.C. under the Zhou Dynasty, the Grand Canal ultimately connected the Yellow and Yangtze Rivers, covering a distance of 2,357 kilometers upon its completion under the Sui Dynasty in the early 600s. Though its completion resulted in greater economic integration of the empire, its construction and maintenance drained the federal treasury and was complicated by harsh working conditions, high material costs, and devastating accidents. The strain of massive building projects such as the Grand Canal, as well as military campaigns, contributed to the dynasty's eventual collapse (Xiong, 2006).

Advances in Breeding and Agricultural Science: Jin through Song Dynasties

In the centuries following the end of Han rule, irrigation projects and efforts to convert saline coastal land to agricultural uses were further expanded (Needham, 1984). During the Northern and Southern Dynasties period, *Jia Sixie*, an agricultural scientist and government official, recorded major advances in agriculture and engineering in his work entitled *Essential Techniques for the Peasantry*. Filling ten volumes (totaling over 100,000 characters), his writings combined a summary of past agricultural knowledge with firsthand observation and experience. The volumes provided officials with a comprehensive guide on agricultural policy and practices, the first of its kind to have survived in its entirety (Needham, 1984).

During the Tang and Song Dynasties, the quality and diversity of crops slowly increased. Farmers began to develop hardier crop varieties by cross-breeding plants that displayed beneficial traits. Breeding was traditionally carried out on the farm and new seed varieties were swapped freely among neighbors. Northern Song Dynasty (960 to 1127) officials promoted rice varieties that ripened earlier in the season, and their widespread introduction enabled farmers to grow multiple crops each year. Bumper rice harvests allowed the country's population to swell to 100 million by the early 1100s, well beyond its previous peak (Ho, 1956). In addition to early ripening, rice varieties were bred for improved drought resistance and suitability for specific uses, such as the production of rice wine. During this period, Chinese science, weaponry, and medicine are thought to have approached the most advanced in the world (Bray, 2000).

New World Crops and Trade: Yuan, Ming, and Qing Dynasties

The development of early ripening rice during the Song Dynasty signaled that Chinese agronomy was in some respects well advanced compared with other parts of the world. During the succeeding Yuan Dynasty, several breakthroughs in cotton cultivation and weaving technologies contributed to the growth of the textile industry. Under the Ming Dynasty, practitioners painstakingly recorded common Chinese crop breeding and cultivation practices, many of which were described in two major texts, the *Exploitation of the Works of Nature* (1637) and *Complete Treatises on Agriculture* (1639). The former was published in Europe and its positive reception prompted its publication in the United States in English under the title *Chinese Technology in the Seventeenth Century* (1966) (Yabuuti, 1967).

Under the Ming Dynasty, China entered a period of little contact with other lands, and its agricultural system developed in isolation. This isolation lasted well into the Qing Dynasty (1644 to 1911), until once again China's economy began to feel the influence of the outside world. As early as the seventeenth century, many crops from the Americas were introduced, including sweet potatoes, corn, and peanuts. Corn in particular could be grown on lands where traditional Chinese crops could not survive, favoring the expansion of cultivation into marginal areas (Needham, 1984).

New cultivation methods widely applied during the Qing Dynasty also helped to improve agricultural productivity. Centuries of experience had helped farmers realize that alternating the type of crops grown on a single piece of land, or allowing land to lie fallow for a few seasons, improved crop yields. This practice, called crop rotation, spread widely under the Qing. Intercropping, the practice of planting several crops together on the same field, also helped to boost land productivity (Needham, 1984). During this period, the government played a key role in the widespread adoption of productivity-enhancing practices and controlling the food supply (Li, 1982). However, productivity gains were still barely keeping up with population growth, and were unable to prevent several devastating famines (Wong, 1982).

Around the same time, China's isolation from the outside world ended as European nations, eager to secure more favorable terms of trade, made several bids for influence in China's port cities. The most famous attempts occurred during the Opium Wars. With foreign exchange resources depleted from continental wars, European buyers resorted to selling opium in return for Chinese spices, silks, and porcelain wares, among other goods. Since the potent drug was banned under the Qing, several clashes between China and Britain (joined by France and the United States in the Second Opium War) over trading rights eventually resulted in the establishment of European economic concessions in the cities of Shanghai and Nanjing, as well as the southern coastal province of Canton (known today by its Chinese name, Guangdong). These developments brought Chinese and Western cultures into closer contact than at any point in living memory. Although some scholars, including China's soon-to-be leader Sun Yatsen, were allowed to study abroad, the Qing leadership suppressed efforts to upgrade the educational system (Hamer & Kung, 1989). The growing foreign presence proved an embarrassment to the Qing government, and contributed to the dynasty's unpopularity around the turn of the twentieth century.

Agriculture and the Environment Prior to 1900

Given the scarcity of arable land, traditional Chinese agriculture relied on many strategies for preserving land productivity. By applying plant material or animal wastes to replenish soils and devising planting strategies to boost yields and reduce pest invasions, farmers were able to realize increases in land productivity. Early Chinese scholarship that described advances in agricultural practices emphasizes soil conservation and other long-term management strategies. Farming techniques included cultivating diverse crops and livestock together in ways that mirrored natural relationships. One such example was the dike-pond system widely used on the Yangtze River Delta. Farmers planted mulberry trees on dikes to feed silkworms, and silkworm feces and discarded pupa were used to feed neighboring pond fish. Pond sludge was then used as manure for the mulberry (Tang, 1984).

However, expansion of farmland and marginal improvements in agricultural technology raised yields, which favored population growth and created a need to expand productivity once again. This accelerating cycle of expansion shaped China's agricultural sector over thousands of years, in spite of setbacks due to drought, floods, disease, and war. The living conditions of most farm laborers did not improve substantially, and many were conscripted to construct massive flood control and irrigation projects. Though China's early advances in agricultural technology allowed for increases in farm output, farming required many hours of intensive labor under harsh conditions, and was fraught with uncertainty and challenges. Furthermore, traditional farming practices were often harsh on the environment. Forests were cleared to expand agricultural land and hillside ecosystems were drastically altered by the introduction of water- and land-intensive paddy agriculture (see Figure 1.5).

Fig. 1.5 Rice terraces in Yunnan province constructed over 1,000 years ago. Photo courtesy of Dr. Chen Zhangliang, China Agricultural University. (A color version of this figure appears between pages 72 and 73.)

Seeds of Change: Fall of the Qing Dynasty, the Republic of China, and World War II

At the dawn of the twentieth century, several developments ushered in a period of dramatic changes in China. The first was the fall of the Qing dynasty in 1911, ending cycles of dynastic rule. The end of Qing leadership followed several decades of growing resentment toward the ruling elite. Clashes with foreign powers had ended in China's relinquishing territory, a source of embarrassment for the incumbent leadership. The transition was accompanied by a growing foreign presence in China and increased exposure to many advances of the industrial age. A wave of intellectual discourse called into question the organization of society and the unequal distribution of wealth, paving the way for social upheaval and power struggles.

China's agricultural sector, shaped by centuries of incremental change and isolation, faced a new world order when it was forced to open its borders in the second half of the nineteenth century. Several industrial revolutions had provided European and American farmers with agricultural technology radically different from that in common use in China. This isolation gradually began to change when Sun Yatsen, a foreign-educated physician and statesman, assumed leadership of the new republic and began to encourage investment in science and technology (Hamer & Kung, 1989). Technologies from abroad began to trickle into China in the 1920s and 1930s as universities were strengthened, scholarly societies formed, the first scientific journals were founded, and a growing number of students were sent abroad for study. During this period, the first modern agricultural schools were founded to teach farming practices and conduct research (Tawney, 1964). This early momentum was lost again at the end of the 1930s with the advent of World War II and civil war.

Despite signs of change, China was still a predominantly agrarian country in the first half of the twentieth century. Most families engaged in agriculture faced significant hardship. Rural households subsisted on meagre diets consisting primarily of staple grains, such as rice or wheat, and rarely consumed meat, vegetables, fruits, and dairy products. The agricultural system was parsed into single-household farms that cultivated their crops on small, often scattered plots of land. With strict controls on international and interregional trade and little support from the crippled Qing and Nationalist governments, new agricultural technologies were slow to emerge or spread (Myers, 1970).

The government of the young Republic of China proved unable to resolve many of the nation's challenges. Opposition from the emerging communist faction further threatened to undermine the shaky republic. The growing rivalry was overshadowed by the need to mount a response to the Japanese invasion during World War II, but civil war erupted again after the Japanese were defeated. Widespread mismanagement of the agricultural sector helped to create favorable conditions for political change and the rise of the Chinese Communist Party. Championing the cause of the oppressed, the communist movement heralded the peasant farmer as its lifeblood and ideal citizen. Support for the movement spread rapidly in a country still nourished by its agrarian roots. Many of the movement's successes rested on promises of farmer empowerment. The inequalities and bourgeois elements of the old order were to be suppressed or expelled. In 1949, the communist faction prevailed and consolidated its power in Beijing, unifying the country and declaring its present-day borders.

Amid the stirrings of foreign presence, revolution, and war, the fundamental organization and technologies on China's farms remained relatively unchanged (Myers, 1970). When the communist government under Mao Zedong assumed power in 1949, China's agricultural sector looked much as it had for centuries. New policies aimed at modernization would have to consider its age-old ingredients— the small family farm, scattered semi-autonomous village communities, and strong local networks of officials and grain merchants. The new government could not overlook the fact that its legitimacy rested on rural stability, as had been true of its predecessors for thousands of years. Against this historical backdrop begins the story of a modern transformation in China's agricultural sector, one that combines the age-old goals of ensuring rural stability and food security with aspirations for global technological and market leadership.

2

Modern Science on the Farm: The Green Revolution

The industrial revolutions of the eighteenth and nineteenth centuries fundamentally altered patterns of daily life and economic activity in many countries. This series of breakthroughs in science and technology, initially concentrated in Western Europe, laid the foundation for the emergence of new industries and many time- and labor-saving inventions. Electrical charges found in nature were harnessed to power offices and homes. Mechanization replaced labor in many industries, driven by savings in cost and time. Steam engines hauled goods across long distances at a fraction of previous travel times, and in the twentieth century, cars emerged as the dominant mode of transportation. Increasingly, societies were prospering economically in proportion to their ability to reap the fruits of scientific and technological advances.

Agriculture also felt the impact of these advances. By the early twentieth century, advances in plant biology were being applied to breed more productive and robust crops. Farmers increased water and fertilizer applications to realize the full potential of high-yield varieties. Synthetic pesticides, many of them highly toxic, often proved more effective than previous methods of pest control. Mechanization reduced the labor requirements of many agricultural processes, such as planting and harvesting.

By the end of World War II, a new balance of power was shaping relations between developed and developing nations. With Nazi Germany defeated, ideological tensions were growing between the United States and the Soviet Union. The Soviet Union began to cultivate communist allies in Eastern Europe and the Far East, while the United States aided reconstruction efforts in Western Europe and Japan. This search for support paralleled accelerating population growth in developing countries, including parts of formerly colonial Africa, Central and South America, and Asia. In some developing countries, particularly those in Asia, nearly all available arable land had been cultivated. Predictions about the ability of nations to meet growing food demand grew increasingly grim. Both the Soviet Union and the United States eyed the situation with growing concern as they sought allies in the developing world.

Against this backdrop, efforts to support agricultural research programs in developing countries found broad support in the United States. Over the following decades, scientific, political, and humanitarian interests combined to aid in establishing modern crop science around the globe. This application of advances in crop

V. J. Karplus and X. W. Deng, *Agricultural Biotechnology in China.*
© Springer 2008

breeding and other associated technologies to address needs in the developing world is now known as the Green Revolution. While many of the breeding techniques were developed in the advanced industrialized countries, international and national centers were set up across the globe to adapt varieties to local growing conditions in developing countries. The successes and failures of the Green Revolution offer several insights for current and future efforts to apply modern science to agricultural challenges.

Scientific Foundations

The world's earliest farmers noticed that, for any particular crop, some individual plants exhibited different characteristics, or traits. Such traits included varied height, stalk strength, or tolerance to pests and drought. For thousands of years, farmers have saved seeds from the most robust plants for replanting the following season, nudging the evolution of major crops in directions that suited the needs and tastes of mankind. Since the world's earliest agricultural civilizations, this practice, known as selective breeding, has become universal in global agricultural systems and has in some cases produced crop species strikingly different from their ancestors. For example, several millennia of corn breeding have yielded modern varieties distinctly different from their wild ancestors in color, size, and taste. Other domesticated staple crops, such as wheat and oilseed rape, have similarly diverged from their predecessors.

The process of selecting robust crops to produce the next generation of seeds was largely hit-or-miss for early farmers, who did not understand how or why the process worked. Nevertheless, selecting seeds solely on the basis of observation produced significant yield increases. As mentioned in Chapter One, early breeding efforts in China produced varieties more suitable for early planting or better able to withstand pest or disease invasions.

From Darwin's Islands to Mendel's Garden

The tide of inventiveness in the 1800s left in its wake new academic disciplines and many unanswered questions. A major dilemma in the emerging discipline of biology was how to explain the great diversity of life on earth. In the 1830s, Charles Darwin traveled to several of the Galapagos Islands off the coast of Ecuador, where he conducted extensive studies on the mating behavior and evolution of finches. He discovered that birds isolated on different islands had evolved into distinct species. This work informed his theory of evolution, in which he posited that the creatures best adapted to a given environment tend to thrive and prolif-erate, while the less fit eventually face extinction. Yet his work was not able to explain exactly how advantageous characteristics were transmitted from parents to offspring.

It was not until the 1850s when the work of a plant scientist and monk, Gregor Mendel, began to reveal this mysterious relationship. By observing the traits of thousands of pea plants over several generations, Mendel discovered that traits disappear and reappear over successive generations in a predictable fashion. Mendel observed that two parents with different forms of a particular trait, such as round or wrinkled peas, produced offspring with only one of the forms, but the "disappearing" form would reappear in the subsequent (third) generation. In an 1866 paper entitled "Experiments on Plant Hybridization," he hypothesized that in order for the numbers to work out right, each parent would have to contribute one trait "factor" to the offspring in either the dominant or recessive form (Mendel, 1866). If a plant receives at least one dominant form of the factor, the plant exhibits the dominant trait. If both forms of the factor are recessive, the plant exhibits the recessive trait.

Mendel's results provided the first evidence that a plant's characteristics are in part determined by those of its parents. His discoveries were overlooked for over 30 years in part because his methods and conclusions were unfamiliar to the scientific mainstream (Dunn, 1965). When rediscovered in the early 1900s, Mendel's work laid the foundation for the new discipline of genetics, named for the "genes" or factors that Mendel had described several decades earlier. Today, we know that patterns of inheritance are in many cases more complex than Mendel described. Still, his work offered the first insights into the basic principles that underpinned age-old breeding practices.

Breeding for Higher Yields

Scientists in the early decades of the twentieth century began to apply Mendel's insights to improve on earlier plant breeding practices. Through carefully designed crosses and close observation of the results, they began to unravel the patterns of inheritance for important traits of the world's major staple crops. Scientists were trained to systematically identify and apply these patterns of inheritance to develop new crops on a scale far greater than was typically performed on a single farm. The ability to cultivate and select from a wide variety of divergent species greatly enhanced the range of potentially successful crosses.

The crop varieties developed during the Green Revolution can be divided into two categories, hybrid and non-hybrid crops, based on the parentage and seed production techniques used to develop them. Hybrid crop seeds are developed by crossing plants selected from substantially dissimilar lines (also known as isogenic lines). These lines are typically different subspecies of the same crop. The main advantage of hybrid crops is their superior yield in the hybrid generation compared with either parent (see Figure 2.1). This property is due to a phenomenon known as "hybrid vigor" or heterosis. While the scientific basis of heterosis remains elusive, its application has become common in breeding programs around the world. Non-hybrid crops, by contrast, are developed by self-pollination (which produces genetically identical plants), sometimes after establishing new traits through cross-breeding among related or dissimilar strains.

Fig. 2.1 Hybrid vigor (heterosis) in corn. The two hybrid corn plants in the center are taller and more productive than both parent corn plants, which were derived from high-quality inbred lines (far left and far right). Reprinted from Birchler, J. A., Auger, D. L., & Riddle, N. C. (2003). In search of the molecular basis of heterosis. *The Plant Cell*, *15*(10), 2236–2239. Copyright 2003 by the American Society of Plant Biologists. Reprinted by permission of the American Society of Plant Biologists via the Copyright Clearance Center. (A color version of this figure appears between pages 72 and 73.)

Since the additional yield or "vigor" of a hybrid crop results from the process of combining the genomes of diverse parents, the newly acquired beneficial properties cannot be uniformly maintained in the next generation by saving seeds produced from hybrid plants. In order to reproduce a hybrid crop, the genetic contributions of the original parents are required. Some correctly point out that farmers who want to retain the benefits of hybrid varieties must repurchase new seed every year instead of reproducing it themselves. In doing so, farmers must consider the value of yield improvements and the price premium among the costs and benefits of adoption.

Unless specified explicitly, varieties mentioned in this book are not hybrids. Many of the early Green Revolution varieties were not hybrids, including wheat strains with shorter, sturdier stature (dwarf varieties) that can support increased grain yield. Non-hybrid crops can typically be reproduced for many generations on farms without compromising yield. Though the ease of breeding crops to express certain traits differs by crop and by trait, desirable traits can be incorporated into either hybrid or non-hybrid crops (see Box 2.1).

Box 2.1 Plant Breeding and Hybrid Varieties: Terminology and Techniques

The choice of a plant breeding method depends on how a particular crop is propagated in the wild. Crop plants typically possess both male and female sexual organs. Some plants reproduce by self-pollination, which involves the production of offspring when a single plant serves as both the male and female parent. Self-pollinating crops include rice, wheat, barley, oats, and edible legumes. Self-pollination (or simply "selfing") produces offspring genetically identical to the parent, also known as pureline varieties. Other crops, such as corn, reproduce by cross-pollination, or the union that occurs when traveling pollen (carrying "male" genetic information) lands on the female organs of neighboring plants. Unlike self-pollination, cross-pollination results in greater genetic diversity among the resulting offspring.

Hybrid crops are the offspring of two genetically pure parents with sufficiently different genetic compositions. The method of obtaining genetically pure ("pureline") parent varieties differs depending on whether a plant is self-pollinating or cross-pollinating. For a self-pollinating variety, breeders must force genetic intermingling by removing a plant's pollen-producing organs and manually introducing pollen from another plant. After a self-pollinating plant is forced to cross-pollinate once, a new genetic combination can be generated and maintained by multiple rounds of self-pollination, yielding a pureline variety. Cross-pollinating plants can also be compelled to self-pollinate until a pureline variety is produced. However, it is more difficult to maintain a genetically stable variety of a cross-pollinating crop in the field because of its tendency to breed readily with neighboring plants.

Hybrid crops do not retain high yield or other beneficial traits after the first generation. This property stems from the dissimilarity of the two parent strains themselves, and thus segregation occurs in subsequent generations. Sometimes the term "hybrid" is mistaken in the popular press to suggest that these crops cannot be reproduced at all. Selfing or cross-breeding a hybrid produces a perfectly viable crop, but without the additional yield gains that result from the combination of a highly diverse parentage.

Once pureline varieties are developed through inbreeding, these varieties can either be crossed with dissimilar relatives to produce hybrids as mentioned above, or crossed with counterparts possessing one or a few desirable traits to transfer it into a superior variety. In the latter case, the introduction of a specific trait is followed by backcrossing, or crossing the offspring with the more robust parent to recover much of the original genetic composition. This process often requires careful observation over successive generations to produce results.

This text draws on information presented by Riley and Constabel (2001).

Global Diffusion of Green Revolution Varieties

Several organizations based in the United States backed initial efforts to apply advances in breeding science to agricultural challenges in the developing world. Breakthroughs in breeding, primarily for corn, had already brought considerable yield gains in parts of the developed world. Many proponents of the technology were hopeful that similar gains could be realized for other crops on a global scale. The Rockefeller Foundation, in cooperation with the Mexican Ministry of Agriculture, established a breeding program known as the Mexican Agricultural Program in 1943. Dr. Norman Borlaug, a freshly minted Ph.D. graduate from the University of Minnesota, was among the first scientists to join the program in 1944. By 1956, Borlaug had developed shorter, sturdier (also known as "dwarf") wheat varieties that could support the weight of larger grain loads. Within a few seasons of widespread planting of the new wheat varieties, Mexico went from importing at least half of its grain to attaining self-sufficiency in grain production (Cleaver, 1972). Although the bilateral Mexican Agricultural Program formally ended in 1962, associated breeding programs were later incorporated into the International Maize and Wheat Improvement Center (CIMMYT) in 1966, which received support from both the Ford and Rockefeller Foundations (Wu & Butz, 2004; Chandler, 1982). In recognition of the substantial impact of his work, Borlaug received the Nobel Peace Prize in 1970.

Meanwhile, researchers at the Ford- and Rockefeller-funded International Rice Research Institute (IRRI) in the Philippines began developing shorter dwarf strains of high-yield rice (Chandler, 1982). Retired scientist Hank Beachell identified the first dwarf rice strain for tropical Asia, IR8, which was developed by crossing varieties from Taiwan and Indonesia. The IR8 strain yielded more and could be planted up to 60 days earlier than conventional varieties. When released across Asia in 1966, the so-called "miracle rice" found markets in India, Burma, Malaysia, the Philippines, and Indonesia (Chandler, 1982). From 1967 to 2001, Gurdev Khush, an Indian scientist at IRRI, developed 300 varieties, several of which became widely planted across Asia (Chandler, 1982). Due in large part to these breeding advances, rice yields across parts of Asia doubled between 1961 and 1990 with only marginal increases in the area of arable land, a feat that earned Beachell and Khush the World Food Prize in 1996 (Wu & Butz, 2004).

Later generations of Green Revolution crops, including many hybrid varieties, diffused worldwide through a network of national and international research and breeding centers. The world's first major hybrid crop, corn, was planted mostly in the United States for several decades before it began to reach farms in developing countries in second half of the twentieth century. China, India, and Latin America adopted high-yield corn varieties that had previously succeeded in raising yields in parts of the developed world. Both wheat and corn yields in Latin America approximately doubled between 1961 and 1991, first in Mexico, then later in South and Central America and the Caribbean (Wu & Butz, 2004). Although many of these gains occurred in China, particularly for rice, collaboration with international centers was not established until 1974 (see Chapter Three for more information).

Greater Outputs, Greater Inputs

Advances in crop breeding were far from the only reason for the yield increases of the Green Revolution. Realizing the harvest potential of high-yielding varieties often required an increase in irrigation, fertilizer, and pesticide applications. Taken together, high-yielding varieties and increases in other inputs helped farmers overcome natural barriers to raising agricultural productivity while reducing the need to expand cultivated land area.

Substantial increases in fertilizer use accompanied the introduction of high-yielding varieties. All living organisms depend on a steady supply of nitrogen, which is an important component of proteins and other fundamental molecules required for life. Although air contains an overwhelming abundance of nitrogen (around 78 percent), it is in a chemical form that plants are not able to use. Plants primarily access nitrogen from the soil in the form of nitrates, which are negatively charged molecules containing nitrogen and oxygen. In nature, a series of soil bacteria convert nitrogen in the air to ammonia and then to nitrates, a process that generates nitrogen compounds more slowly than most crops can consume them. Shortages grow especially acute if nitrogen is not replenished after repeated plantings, since ammonia and nitrates are easily depleted from soils by erosion or water runoff streams (Smil, 2004; Matson, Parton, Power, & Swift, 1997). Maximizing the yield potential of high-yielding crop varieties nearly always requires supplementing natural conditions with artificial nitrogen sources, which until the early twentieth century were not widely available.

Nineteenth century scientists understood that a lack of readily available nitrogen was limiting agricultural production. This problem was more difficult to resolve for nitrogen than for other nutrients, such as phosphorus or potassium, which could be mined and applied directly. Breakthroughs by Carl Bosch at the German chemical company BASF and Fritz Haber at the Technical University of Karlsruhe in Germany provided the technical basis for a method of converting gaseous nitrogen to ammonia, a chemically active nitrogen-containing compound (Smil, 2004). Instead of relying solely on natural sources, ammonia could be synthesized in great quantities and added to soils. Ammonia synthesis had many other applications, most notably in the production of explosives for the Great War (World War I), which was just beginning as the first ammonia plant started production in 1913 (Smil, 2004). After the end of World War II, production shifted to meet the growing demand for fertilizers, which helped to support the broader adoption of high-yielding varieties. By the turn of the twenty-first century, the world's fertilizer plants were producing 80 million tons per year using the highly energy-intensive Haber-Bosch process. Fertilizers are now routinely sold and applied on a massive scale around the world. Some estimates suggest that artificial nitrogen sources supported the equivalent of around one-third of the earth's population by the late 1990s (Smil, 1997).

High-yielding varieties are also on average thirstier than their predecessors, making access to water through irrigation systems or natural rainfall particularly important. As a result, irrigated land area worldwide increased almost 1.7 times over the last half century, and in developing countries, irrigated land approximately doubled from 100 million to 200 million hectares (Tilman, 1999). By the early twentieth century, around 17 percent of all agricultural lands were irrigated, but these lands accounted for 40 percent of the world's food supply (Pimentel, 2004). Today, agriculture places the largest strain on fresh water supplies, consuming approximately 70 percent of total fresh water withdrawn every year (Pimentel, 2004). Though gains vary by crop and region, a significant percentage of Green Revolution yield gains can be attributed to increased water use, but at the price of land degradation and depletion of local water supplies over the long term (Ruttan, 2002).

Post-industrial advances in chemistry paved the way for the development of chemical pesticides, herbicides, and fungicides, many of which proved more effective than the methods they replaced. However, some chemicals used in the world's first synthetic pesticides were later discovered to have toxic properties largely overlooked by the first users. As adoption proceeded, activists and environmentalists led an outcry against particularly toxic forms. Perhaps the most famous example is the pesticide DDT condemned in Rachel Carson's popular 1962 book *Silent Spring* (Carson, 1962). Large chemical companies were drawn into the agricultural sector on an unprecedented scale to produce pesticides and other inputs.

Although the use of machinery in agricultural production proved economically attractive for many farms in the developed world throughout the twentieth century, it did not rank among the major advances of the Green Revolution. Explanations for the discrepancy point out that technology choices depend on a country's endowment of land and labor, as well as their relative costs (Ruttan, 2002). At the beginning of the Green Revolution, many countries in Asia and Latin America had an abundant supply of agricultural laborers with few other employment options. As a result, labor-saving technologies, such as those typical of mechanized agriculture, were adopted more slowly.

Successes of the Green Revolution

Few question that the Green Revolution helped to avert widespread food shortages in many parts of the world, most notably in India and other parts of Southeast Asia. Its greatest successes were perhaps the widespread yield gains that allowed food supply to keep pace with demand. These gains, particularly for rice, are thought to have laid the foundation for the recent economic booms in Southeast Asia and current rapid growth in India and China (Hu, Huang, Jin, & Rozelle, 2000). Yet the Green Revolution did not spread by itself—it was enabled by farmers as well as international and local organizations.

The successes of the Green Revolution, which began with the wheat breeding efforts of Borlaug and others in Mexico, were only realized on a global scale with support from a global network of national and international agricultural research centers (NARCs and IARCs). The network grew out of successful models in Mexico (CIMMYT) and the Philippines (IRRI) to include additional IARCs focused on specific crops, cropping techniques, and safety issues. In 1971, these centers were linked in a global network as part of a World Bank initiative to establish the

Consultative Group for International Agricultural Research (CGIAR). These centers involved local scientists and focused on introducing desirable traits, mostly into indigenous varieties, to overcome local agricultural challenges. As research centers that focus on in-country challenges, NARCs provided additional support and a critical link between international sources of technology and local farmers (Evenson, 2004). Underwritten by donor support from governments and private foundations, the CGIAR network was a critical enabler of the Green Revolution in many parts of the developing world. Today, its centers continue to develop and introduce new varieties and train farmers, breeders, and scientists around the globe.

Relatively free access to advances generated by publicly-funded agricultural research centers helped the diffusion of the Green Revolution varieties and associated agricultural inputs (Evenson & Gollin, 2003). Much of the research was conducted in publicly funded institutes, which meant that access to advances made through research was not restricted by strong intellectual property rights. Instead, universities as well as national and international research institutes were able to share the tools needed to develop new varieties adapted to local conditions, considerably reducing costs and improving access to necessary techniques (Wu & Butz, 2004). At the farm level, the wheat and rice varieties developed during the Green Revolution spread rapidly due to farmer-to-farmer exchange following introduction.

Global consensus on the urgency and importance of shoring up food security also helped to reduce barriers to diffusion of the new agricultural technologies. Averting widespread famine and stabilizing emerging economies in Latin America and Asia were the primary rationale used to mobilize support for the programs (Wu & Butz, 2004). Lack of an organized, widespread opposition to the adoption of these varieties further helped to smooth the road for investments in breeding research programs.

Shortcomings and Remaining Challenges

Despite considerable measurable benefits, the impact of the Green Revolution was neither universal nor entirely positive. Perhaps the harshest critique of the Green Revolution is that it bypassed large swaths of the globe, particularly in Africa, where diets include very few of the crops targeted by the initial Green Revolution efforts. Many crops developed were not suited to the thinner African top soils and long, unpredictable periods of drought. Poor infrastructure also made it difficult and expensive to deliver the necessary inputs to maximize yield gains. African crops have long suffered from problems unique to the region. For example, many of the pests, weeds, and diseases that affect important crops such as corn, cassava, and bananas in Africa are distinct from those found in other parts of the world (Thompson, 2002). Only in the last twenty years have efforts at several of the Consultative Group centers, as well as by the Rockefeller Foundation and other organizations, focused on developing improved crop varieties better suited to African agriculture and diets (Rockefeller Foundation, 2006).

The introduction of some varieties resulted in changes to local farming systems as new technologies mixed with or replaced traditional practices. Since hybrid seeds lose their added value after the first generation and farmers do not possess the specially bred parent plants required to reproduce seeds on farms, farmers need to purchase new seed each year. Hybrid seed use also requires seasonal purchases of inputs, which can be costly and cumbersome to transport, especially if road conditions are poor and farms are located far from suppliers.

Environmental degradation has also grown more acute as a result of the introduction of high-yielding varieties and associated inputs, which include irrigation, fertilizers, and pesticides. Chemicals used in fertilizers, herbicides and pesticides have taken a serious toll on the surrounding land, water, and public health. Nitrogen- and phosphate-based fertilizers have accumulated in surface water supplies, altering resident ecosystems in nearby lakes and streams (Smil, 1997). Fertilizer production requires high energy inputs, which are often supplied by increasingly scarce and pollution-intensive fossil fuels. Increases in water use have also caused shortages and related problems. Over-irrigation of crop lands has leached soil nutrients and increased soil salinity, leaving fields unable to support crops. Meanwhile, water is growing increasingly scarce in some arid regions, as many underground aquifers are being depleted faster than they are being replenished (Ruttan, 2002).

Others question the nutritional value of high-yield varieties, which are often bred to increase harvests, delay ripening, and contain other qualities desired by farmers and processing facilities. A recent study described how breeding for these so-called "input traits" may have inadvertently reduced the expression of other "output traits" that are important to consumers, such as taste or cooking quality (Morris & Sands, 2006). Some consumers are concerned that high-yield varieties are less flavorful than their predecessors, and point out that the trend away from catering foremost to consumer preferences is likely to continue in future generations of the technology. In China, for instance, consumers have previously complained that some hybrid rice varieties were not very flavorful.

The limitations of Green Revolution technologies now rank among the major challenges facing agricultural policymakers, farmers, breeders, and scientists today. Nations that reaped many of the Green Revolution's benefits are now seeking solutions for (or ways to avoid) land degradation and a slowdown in agricultural yield growth. Meanwhile, the Rockefeller Foundation and other supporters of the Green Revolution are refocusing their efforts on ways to reach populations that did not initially benefit from the past introduction of agricultural technologies, particularly in Africa. If past experience is any indication, the policy environment—both global and local—is likely to be at least as important as the available scientific tools in tackling this next generation of challenges.

3

Transformation in China's Agriculture in the Twentieth Century

Against the backdrop of world conflicts during the first half of the twentieth century, China experienced several decades of internal upheaval. The fall of the Qing Dynasty, war with Japan, and civil war disrupted lives, harvests, and commerce. When the dust began to settle, China's economy had suffered significant losses. By the time Mao's government took power in 1949, production had decreased 25 percent in agriculture, 30 percent in light industry, and 70 percent in heavy industry compared to pre-war levels (Huang, 1998).

The first priority of China's new leaders was to launch an ambitious industrialization program based on the Soviet model. Agricultural policy was formulated with the needs of heavy industry in mind. This practice lasted into the mid-1970s, when Mao's death and the policy shift that followed ushered in sweeping economic reforms, a slow opening to the world economy, and increased emphasis on science and technology as a means to achieve national development goals. It was against this historical backdrop that many advances in breeding technology developed and spread widely to China's countryside.

This chapter begins by chronicling the changes in agricultural and industrial policies from 1949 through the present that influenced the development and adoption of advanced breeding methods in China, including many of the same techniques that underpinned the Green Revolution (see Figure 3.1). Our focus then shifts to the development and adoption of specific inputs, including seeds, irrigation, and fertilizers. The successes and failures of these technologies and practices in China are then considered in a discussion of the future challenges facing the country's agricultural sector.

Collectivization and Industrialization: 1950–1961

In the early 1950s, central government officials promoting industrialization could not ignore the fact that China was still a largely agrarian nation. The vast majority (88 percent) of the nation's population lived and worked in rural areas (Huang, 1998). In 1952, per capita grain output could barely support the population, and

V. J. Karplus and X. W. Deng, *Agricultural Biotechnology in China.*
© Springer 2008

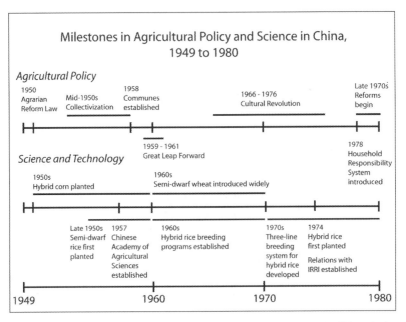

Fig. 3.1 Overview of agricultural policy and technology development in China during the pre-reform period.

China had little industrial capacity outside of a few major cities (Huang, 1998). In contrast to the Soviet Union, with its highly centralized leadership and high proportion of urban residents, China's new government had to elicit the cooperation of a vast and diffuse rural population. As a result, many policies to promote industrialization adapted from the Soviet Union had unique character and outcomes in China.

In the early 1950s, land ownership reform was at the top of the government agenda, both to equalize classes according to communist ideology and consolidate power for a national industrialization program. The Agrarian Reform Law of 1950 was the first in a series of national policies to eliminate private ownership and redistribute land to farmers. Retribution against longstanding inequality accompanied the process of redistribution. Land ownership was revoked, and many landlords and rich farmers were designated class enemies and often brutally displaced. The reforms proved popular among the country's many smallholder farmers, helping the new government to consolidate power.

The gradual redistribution of agricultural land into communes, or large, collectively-owned farming organizations, proceeded according to the Soviet model. The astonishing output levels (later recognized as fabricated) at communes in the northeastern Soviet Union convinced China's leadership that similar gains could be achieved within its own borders. During the 1950s, households were consolidated in stages into collectives of fewer than 200 families, culminating in the establishment of much larger units known as communes in 1958. Communes included around 5,400 households (20,000 to 30,000 members) and were organized into three administrative levels: communes, brigades, and production teams. Farmers were told which crops to plant and how much of harvests had to be supplied to state procurement organizations according to strict quotas. Production income was divided among members of the collective based on a combination of points earned for work performed and the principle of equal distribution (Domes, 1980).

Collectivization was just one of many measures taken by Mao's government to lay the foundation for an ambitious industrialization campaign. Other sectors, including agriculture, fell second to the all-important cause of raising iron and steel output. Food prices were kept as low as possible to minimize costs to industries and urban dwellers without entirely robbing farmers of incentives to produce. Industries were nationalized to ensure that any profits were recovered by the government or directly reinvested in the industrialization effort. Interest rates were kept artificially low to stimulate investment, while some 156 key projects were announced to upgrade production capacity, and entailed large purchases of Soviet technology (Huang, 1998). Trade with the outside world slowed to a trickle as the government pursued a policy of economic and food self-sufficiency. Exceptions were made only for limited exchanges with other communist countries. Agricultural exports, mostly to the Soviet Union, quickly became a precious source of foreign exchange to purchase improved technology. Meanwhile, Chairman Mao encouraged families to have more children to strengthen the country's labor force.

The Great Leap Forward and Cultural Revolution: 1959–1976

To invigorate China's comparatively low industrial capacity, in the late 1950s Mao announced the Great Leap Forward, a program that enlisted China's considerable manpower in a backyard steel production effort. Farmers were expected to melt down all available metal objects, including essential farm tools, household utensils, and furniture, in makeshift backyard ovens. Though in theory this effort was supposed to capitalize on China's abundance of labor, the program failed, not least because the steel on most farms was of very poor quality and unsuitable for large scale industrial projects. It also detracted from the time farmers spent in fields, which was reflected in sharp reductions in yields.

The implementation of the commune system in 1958 and the Great Leap Forward preceded one of the most catastrophic famines in recorded history. From 1953 to 1957 (China's First Five-Year Plan), grain production initially increased steadily, but then stagnated and had dropped significantly by 1960 (Domes, 1980). Official reports of grain harvests were highly exaggerated, in some localities by a factor of two or more (Becker, 1998). Initially, leaders heralded the initiatives as a great success. Photographs in the *China Pictorial* showed children atop wheat harvests so thick they could stand on them, but the images were later found to be staged by placing a bench under the children (Economy, 2005). Although estimates of the famine's death toll range widely, studies based on single-year age distributions from the 1982 census placed the total between 16.5 and 30 million (Smil, 2004). The ill-conceived policies of the Great Leap Forward and unusually harsh weather conditions are often cited as primary causes. Official reports attributed the famine to 70 percent human error and 30 percent natural disaster (Becker, 1998).

The precipitous drop in agricultural output has been attributed primarily to the lack of individual work incentives under the commune system (Kung & Lin, 2003). Under this system, farmers lost all claim to the land they worked and their free daily food ration only loosely, if at all, reflected production effort. Often yields were inflated, as production leaders sought to gain favor with party superiors. This situation left farmers with few reasons to care adequately for their crops and land or make careful planting decisions. Instead, instructions handed down through the commune administration mandated the type of crops to be planted and the timing of sowing and harvesting.

An inward-looking trade policy in part helped to sustain the Chinese industrial experiment because nationalized industries faced no competition from foreign businesses, nor did they have to respond to global market prices. Though this allowed inefficiencies to persist, the output levels of China's fledgling heavy industries grew almost sevenfold between 1952 and 1978 (Huang, 1998). At times, this growth came at the expense of agriculture and food consumption. During the worst famine years, the government continued to promote agricultural exports in order to maintain precious foreign exchange reserves. From 1957 to 1961, nearly 12 million tons of grain and large amounts of cotton yarn, cloth, pork, poultry, and fruit were exported even as a large percentage of the population was starving (Becker, 1998).

In the early 1960s, the failures of the Great Leap Forward prompted the dismantling and reorganization of communes into production teams consisting of 20 to 30 households. Production teams assumed far greater responsibility for making planting decisions, and markets for crops other than grains were allowed to develop on a limited basis. Even leaders and their families in party headquarters in Beijing began to cultivate food crops in backyard gardens (Becker, 1998). National leaders introduced several reforms to promote the rural economy, including measures that allowed farmers to cultivate crops on small parcels of unused land and sell some non-grain crops (Becker, 1998). High-yield seeds were distributed, and irrigation facilities were repaired and expanded.

Although a commitment to scientific advance persisted in official rhetoric, China in the 1950s still remained far removed from academic communities in other parts of the world. Official dismissal or downplaying of established scientific knowledge only widened the gap. During the Great Leap Forward, many agricultural advisors rejected basic tenets of physics, chemistry, and biology. Some made bogus claims, many of them based on the specious science of Soviet agricultural advisor Trofim Lysenko. News reports lauded the creation of giant vegetables and "red cotton" by cross-breeding cotton and tomato (Becker, 1998). Another campaign to raise agricultural yields called upon citizens to eliminate flies, mosquitos, rats, and sparrows (Economy, 2005). Although many "Lysenkoists" were replaced with accomplished practitioners in the early 1960s, a research tradition that promoted independent scientific inquiry had yet to take root.

Encouraging trends reversed again during the Cultural Revolution (1966 to 1976). The central government mandated the relocation of universities and work units to the countryside, dividing families and forcing many to labor at jobs for which they were ill-prepared. Students were encouraged to rebel against teachers and parents, and colleagues against each other, leading to widespread distrust and destruction. Intellectuals were downgraded to the "stinking ninth" class. Very few institutes were able to continue work during these years. A notable exception was the Shanghai Institute of Biochemistry, which during this period became the world's first to synthesize bovine insulin (Hamer & Kung, 1989; Cao, 2004). Work was also allowed to continue half-time at the Chinese Academy of Sciences. However, overall industrial production and education suffered greatly, while low levels of per capita food consumption remained unchanged from 1957 to 1978 (Smil, 2004).

Signs of Change: 1978 to the Present

The excesses of the Cultural Revolution did not go unnoticed by a new generation of emerging leaders. Mao's death and the dismissal of the Cultural Revolution's other chief proponents paved the way for a new wave of policy experimentation at the end of the 1970s. Led by Deng Xiaoping, in 1978 China's government embarked on a sweeping reform program aimed at economic modernization. Guiding the reforms was a newfound emphasis on science and the market. Universities opened their doors again and reinstated many professors and administrators (Becker, 1998). China also began to open its economy to the outside world. Many scholars were sent abroad, billions of dollars in foreign technology were imported to modernize the economy, and competition gradually emerged in many sectors (Suttmeier & Cao, 1999; Oldham, 1997).

The agricultural sector was the first to undergo reforms, starting with changes in land tenure and increases in agricultural procurement prices. Beginning in 1978, land reforms began with small-scale experiments that rewarded farmers according to the harvest they produced on a rented parcel of publicly-owned land. Although first undertaken with caution on a limited scale, initial experiments with household-based incentives were so successful in increasing production that the program quickly expanded to other parts of the country (Lin, 1988). Known as the Household Responsibility System (HRS), this arrangement offered long-term leases to farmers for a period of fifteen to thirty years. Farmers were also allowed greater autonomy in their planting decisions (Carter, Zhong, & Cai, 1996). Some farmers had experimented with household-based production incentives prior to the establishment of communes, perhaps one factor that contributed to the rapid adoption of the new system (Huang, 1998). Early measures to improve production incentives included raising procurement prices for some goods, although these measures were not originally intended to foster the emergence of markets (de Brauw, Huang, & Rozelle, 2002).

Prior to the reforms, agricultural goods were divided into three categories, as described by Carter et al. (1996). Staple grains, cotton, and oil-bearing crops were classified as "unified goods" and acquired by the state according to fixed prices and mandatory quotas. "Dual track" goods included meat, fish, tobacco, and tea, which could be sold at market prices once quotas had been fulfilled. Fruits and vegetables fell into a third category, "zero quota" commodities, that were typically not subject to quotas, but in the absence of a market, sales were still controlled by the state. During the 1980s, quota restrictions were loosened, and government contract purchasing was reduced, encouraging the formation of nascent markets (Carter et al., 1996).

During the 1980s and 1990s, additional reforms created incentives for local officials, banks, and businesses to encourage the growth of rural industries (Oi, 1999). Reforms that allowed township and village enterprises to keep or reinvest profits above certain levels strengthened business incentives for local entrepreneurs. Rural enterprises, old and new, became more profit oriented, and absorbed some of the excess labor from the countryside (Economy, 2005). Local banks, businesses and credit cooperatives helped to support growth with lenient lending policies (Oi, 1999). Liberalization of prices and quotas favored entrepreneurial farmers, who began to open small businesses using surplus earnings. However, many farmers continued to hold on to their land rather than allow it to be consolidated with neighboring farms. Land still provided security when farmers were faced with uncertain and rapidly changing central economic policies.

Taken together, the reforms had a substantial impact on both the productivity of the agricultural sector and rural livelihoods. Long-term leases gave farmers a renewed interest in maintaining their land and increasing crop yields. The prices at which farmers sold their crops were allowed to rise. Planting decisions were increasingly determined by farmers rather than government planners, allowing farmers to consider which crops were truly best suited to their land and household budget each year. In Shandong province, farmers reverted to planting cotton, which was better suited to the land, once the government's grain quotas had been relaxed (Crook, 1988). Rural industries also absorbed excess labor from agriculture, providing an attractive alternative to augment or supplant farm incomes. The impact on rural welfare was also substantial (Lin, 1988). During the period from 1978 to 1985, one quarter of the rural population emerged from below the absolute poverty line, as the number of people in this category fell from 270 million to 100 million (Piazza & Liang, 1998).

Yet the impact of reforms was not entirely positive. The breakup of collectives resulted in a decline in investment in public infrastructure, such as irrigation. Townships and villages, which inherited some of the functions of communes and production brigades, were frequently unable to fill this gap (Huang, 1998). As planting decisions shifted in favor of more lucrative or geographically suitable crops, the planted area of some crops—particularly staple grains—began to fall. Record grain imports from abroad in the mid-1990s prompted speculation that China's agricultural sector would be unable to meet rising food demand (Brown, 1995). The government reinstated quotas for many crops amid evidence that China's grain self-sufficiency was eroding (Carter et al., 1996). Widespread migration resulting from the growth of both rural and urban economic opportunities uprooted lives and placed increasing burdens on urban infrastructure and social services. Today, migrants comprise between 10 and 33 percent of the total population in many cities, a trend that parallels a decline in the total agricultural workforce in China from 70 percent of the employed population in 1978 to around 40 percent in 2005 (Economy, 2005; *China Statistical Yearbook,* 2006). The reforms also included a reduction in state support for research institutes. The impact on the agricultural research system will be discussed later in this chapter.

Agricultural Science and Technology

Challenges facing the agricultural sector provided an important catalyst for domestic investment in agricultural research in the 1950s and 1960s. In the decades leading up to 1949, shortages of fertile land were already limiting the ability of China's agricultural sector to provide for the country's growing population. In the early to mid-1900s, China's population had begun to grow at an unprecedented rate. State policies encouraging large families, in part to increase tax revenues and fill armies, only reinforced this trend. Despite many deaths in the famine of the early 1960s, the population continued to swell in subsequent decades, prompting Deng Xiaoping to announce the One Child Policy in 1979. Yet limiting Chinese couples in urban areas to a single child (exceptions were made in rural areas and for ethnic minority groups) did little to turn the tide in the near term, as momentum had already been established in earlier years.

The broader policy developments described above paralleled—and in many cases, heavily influenced—the emergence of China's present agricultural research system. During the pre-reform period, research was limited by low funding and resource diversions to industrialization. Nevertheless, policies explicitly invoked science and technology as part of the solution to raising agricultural production (Becker, 1998). While at times government-backed scientists supported technologies that later proved to be ill-conceived, support for agricultural research suggested a strong belief in the potential of science and technology to improve productivity. Throughout the second half of the twentieth century, China's agricultural research system played a major role in developing improved seed and other inputs suited to local agricultural needs. Although the contribution of international centers increased in the 1970s, the strengths of China's domestic research programs played an essential role in raising the country's agricultural productivity and enabling collaboration efforts.

China's Agricultural Research System

The creation of an extensive agricultural research system underpinned the development and diffusion of modern agricultural technologies in China (see Figure 3.2). Capacity for agricultural research was established at nearly every level of government. The Chinese Academy of Agricultural Sciences (CAAS) was established in 1957 as the principal research branch of the Ministry of Agriculture (Tang, 1984). The agricultural school system, which had been established in the late nineteenth and early twentieth centuries, was drastically restructured and many schools were consolidated and designated "agricultural colleges" or research institutes. Subordinate to the Ministry of Agriculture, these institutes laid the foundation for China's agricultural research system. Throughout the twentieth century, they provided an important channel for the introduction of agricultural technologies. The agricultural research system remained distinct from the Chinese Academy of Sciences, which housed fundamental research in the natural and social sciences and will be discussed later in Chapter Five.

Economists have reported that the research system contributed significantly to increases in agricultural productivity before and during the reform period (Hu, Huang, Jin, & Rozelle, 2000). The major gains in crop production in China came from the introduction of high-yielding varieties, whose productive potential was enhanced by improving water control and increasing fertilizer usage (Stone, 1990). Although the introduction of the Household Responsibility System provided a sizeable one-time boost to agricultural production, a large share of gains in agricultural productivity during the 1970s and 1980s can be attributed predominantly to the introduction of new technology (Huang & Rozelle, 1996).

Fig. 3.2 Mural in the entrance hall of the China Agriculture Museum, Beijing, depicting the role of advances in crop genetics and breeding in China's agricultural development. Photo by authors. (A color version of this figure appears between pages 72 and 73.)

High-Yielding Varieties

Improved crop varieties developed through breeding programs were the top contributor to the expansion of China's agricultural output during the later half of the twentieth century (Huang, Hu, & Rozelle, 2002b). Beginning in the early 1950s, China's agricultural research system benefited from ambitious programs to acquire superior varieties from abroad, including cotton, corn and sorghum varieties (Pray, 2001a). These varieties were then disseminated to research institutes for the breeding of varieties suited to local conditions (Stone, 1990). This process benefited from considerable local expertise and an already high number of available varieties. Most of the new varieties were developed through research efforts in Chinese institutes, using techniques developed in China and abroad. These early efforts to develop breeding science expanded to include a national network for breeding, testing, producing, and disseminating high-yielding varieties, with the agricultural research institutes at the national, provincial and prefectural levels as its backbone.

Rice was perhaps the most important crop in China to be affected by advances in breeding techniques. China's southern region has long produced the greatest share of the country's rice supply, so it is unsurprising that rice breeding programs first emerged in the south. First developed in 1956, a rice dwarf variety introduced in China was bred to support heavier grain yield and to mature earlier than normal (Athwal, 1971). Dwarf varieties were bred for responsiveness to fertilizer applications, which helped to boost plant growth. The widespread planting of these varieties, along with improvements in irrigation, helped to usher in a wave of yield gains in China that coincided with the Green Revolution.

Fig. 3.3 High-yielding hybrid "super rice" growing in Hunan province. Photo courtesy of Yang Yuanzhu, Yahua Seed Corporation, Changsha, Hunan, China. (A color version of this figure appears between pages 72 and 73.)

The second breakthrough came in the 1970s, when the CAAS and the Hunan Academy of Agricultural Sciences developed the "three-line" system of hybrid rice production (led by Chinese scientist and breeder Yuan Longping, see Box 3.1). Later in the 1990s, Chinese scientists introduced the "two-line" system of hybrid breeding, which significantly reduced the time required to generate suitable plant material for seed production. The development of the "two-line" system was made possible by new insights into the genetic underpinnings of male sterility in plants (see Box 3.2 for more information). The introduction of hybrid rice was in part responsible for the 2.3 percent increase in rice yield frontiers (defined by the maximum rice yield observed in a given province and year) across China from 1980 to 1995 (Hu et al., 2000). A high rate of turnover of rice varieties (up to 20 to 30 percent) further indicates the success of China's research program in providing improved varieties (Hu et al., 2000).

Box 3.1 The Farmer-Breeder Yuan Longping and Hybrid Rice

Born into a rural farm household in 1930 in Qianyang, Hunan province, Yuan Longping has dedicated his life to breeding superior rice varieties. After graduating from the Southwest Agricultural College in 1953, Yuan began a career as a teacher, before later devoting himself full-time to the science of crop breeding. In 1964, Yuan improved on rice cytoplasmic male sterile strains for breeding purposes, and developed the "three-line" breeding system for hybrid rice. Hybrid rice is now widely grown across southern, central, and coastal regions of China, the main rice-growing regions. Yuan is credited as the first to develop hybrid rice, an achievement that earned him the title of "Father of Hybrid Rice."

Most of Yuan's work in recent decades has been conducted at the Hunan Hybrid Rice Research Center (HHRRC) in Changsha, Hunan province. The HHRRC has led the development of the three-line system of hybrid rice breeding. In connection with this program, Yuan produced a commercial variety of hybrid rice known as *nanyou* No. 2, which was released in 1974 and yielded 20 percent more than the varieties it replaced. Hybrid rice varieties have since spread to well over half of China's rice planting area. Total area cultivated with hybrid rice expanded over threefold between 1978 and 1990, from 4.3 million hectares to 15.9 million hectares, with particularly large gains in poor inland regions (Hu et al., 2000). The yield boost is estimated to be enough to feed an additional 60 million people and was the primary accomplishment that won Yuan the World Food Prize in 1996. Yuan was also awarded China's State Supreme Science and Technology Award, also known as China's Nobel Prize.

In recent years, Yuan has been developing and field testing an improved strain known as super hybrid rice, which yields 30 percent more grain than earlier hybrid varieties. The super hybrids are tall and straight with low-hanging panicles that support the enlarged rice grains (see Figure 3.3).

Fig. 3.4 Yuan Longping (left) with Yang Yuanzhu (right), a leading breeder from the younger generation, in a field of super hybrid rice in Hunan province. Photo courtesy of Yang Yuanzhu, Yahua Seed Corporation, Changsha, Hunan, China. (A color version of this figure appears between pages 72 and 73.)

Wheat, a staple crop grown primarily in northern China, was another important target of China's breeding programs. The first semi-dwarf wheat varieties in China were developed by Chinese farmers in the south prior to the 1950s. Research and development of semi-dwarf varieties formally began in the 1950s, and many varieties were introduced widely over the following decades (He, Rajaram, Xin, & Huang, 2001). In particular, Professor Zhuang Qiaosheng began a long and productive career in developing high-yielding wheat varieties during this period. One of China's most recognized wheat breeders, Zhuang worked for many years at the CAAS Institute of Crop Breeding and Cultivation, and was later selected as an academician of the Chinese Academy of Sciences.

In addition, rust-resistant wheat varieties were developed during the early 1950s and planted widely. Many of these gains were due to the pioneering work of Professor Li Zhensheng, a geneticist and wheat breeder at the CAS Institute of Genetics and Developmental Biology. By breeding domesticated wheat with wild varieties, he developed wheat strains better able to withstand outbreaks of stripe rust, a fungal disease. He was later instrumental in advising government efforts to increase grain yields in the early 1990s. Also an academician of the CAS, Li was awarded in 2007 the Supreme Award of Science and Technology in China for his contributions.

Box 3.2 Hybrid Rice: The Two-line and Three-line Hybrid Breeding Systems

Of the methods of breeding hybrid rice in widespread use today, most were pioneered or advanced in China over the last half century. Hybrid rice results from the cross-breeding of two genetically dissimilar parents. The two parent rice plants typically belong to different sub-species. Since hybrid vigor results from the genetic disparities between the two parents, its presentation in offspring is limited by the extent of the parents' diversity.

Rice is a self-pollinated crop. As a result, cross-breeding requires that the male organs of at least one of the plants be sterile so that the plant can be pollinated by a genetically dissimilar variety. This problem does not arise in the case of crops that naturally reproduce by cross pollination (see Box 2.1).

Methods for breeding hybrid rice plants include the three-line and two-line systems. The three-line system is the oldest system and is more time and labor intensive, mostly because it requires creation of three separate lines. A male sterile plant (A – see Figure 3.5) is crossed with a maintainer line (B) that is genetically identical to the A line, except that it produces pollen and enables the propagation of the sterile line. The male sterile rice plants are then allowed to cross with a genetically distinct restorer line (R), resulting in the production of fertile hybrid seeds.

In the two-line system, by contrast, one parent is a conditional male sterile, allowing plants to self-propagate under one growth condition, while they are rendered sterile under another growth condition. While a plant is male sterile, the female organs of the plant can be dusted with pollen from the other parent. Inducing plants to switch from fertile to male sterile can be accomplished in several ways. One method employs genes that turn male sterility on or off in response to environmental cues, such as temperature or day length. For instance, a plant would be able to self-pollinate at low temperature, but at high temperature it would be rendered male sterile, allowing it to reproduce using pollen from a genetically dissimilar species. Another method of rendering one parent sterile involves applying chemicals directly to disable its pollen-producing organs. This method has reportedly been used commercially in China until it was discontinued due to the health risks of applying the chemicals, which included arsenic-containing compounds (Virmani, Sun, Mou, Ali, & Mao, 2003). See Figure 3.5 for a comparison of the advantages and disadvantages of the three methods.

Many scientists in China and elsewhere are focusing on identifying strategies for overcoming known yield limitations. China's super hybrid rice program has focused on creating varieties by crossing one indica strain with another rice variety derived from a combination of indica and japonica sub-species, thereby exploiting the greater genetic diversity between the parents. As scientists gain access to advanced tools for predicting the outcomes of various genetic combinations, they will be able to more precisely design crosses that can be expected to maximize yields.

Hybrid Rice Breeding Systems

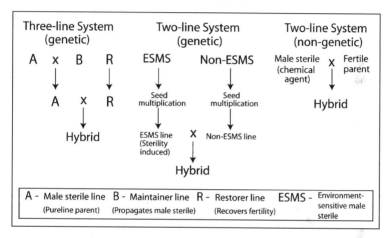

The Two-Line and Three-Line Hybrid Breeding Systems Compared

Three-line System (genetic)	Two-line System (genetic)	Two-line System (non-genetic)
Advantages	**Advantages**	**Advantages**
- Not constrained by environmental factors - High purity of hybrids	- Requires less time than the three-line system - EGMS trait easy to introduce - No need for restorer genes, which some varieties lack	- Wide range of varieties can be used - If chemical treatment fails, no loss occurs since female can still self-pollinate
Disadvantages	**Disadvantages**	**Disadvantages**
- Time consuming compared to other methods - Requires more plant materials	- Sterility requires reliable environmental conditions	- Impure hybrids result if chemical treatment incomplete (for instance, due to weather) - Some chemicals used are toxic - Chemicals can be expensive

Fig. 3.5 The two-line and three-line hybrid rice breeding systems. Partially adapted from Virmani, S. S., Sun, Z. X., Mou, T. M., Ali, A. J., & Mao, C. X. (2003). *Two-line hybrid rice breeding manual*. Los Baños, Laguna, Philippines: International Rice Research Institute.

Hybrid wheat breeding programs officially began in China in 1965, but their impact has been limited. Except for the famine years (1960 to 1962), wheat yields in China increased between 1958 and 1976, in large part due to the work of provincial- and county-level plant breeders (Stone, 1990). In the mid-1980s, breeding programs also began to focus on quality traits, such as higher gluten content to improve the consistency of foods, such as noodles. Efforts to develop improved wheat varieties have benefited as much from the contributions of local breeders as from those of scientists in national research programs.

Other crops besides wheat showed impressive gains as well. During this period, Chinese-bred varieties of sweet potato, corn, and sorghum were yielding as much as the best elite varieties anywhere in the world, and surpassed the yields achieved by prominent crop breeders abroad (Stone, 1990). Together with scientists at the International Potato Center in Peru, Chinese scientists also developed a virus-free potato strain in the late 1980s, which reduced a persistent cause of poor yields (Fuglie, Zhang, Salazar, & Walker, 1999).

Corn yields also benefited from the introduction of hybrids as early as the 1950s, and the range of hybrids increased throughout the 1970s. Corn is grown primarily in the northern part of China, supplementing wheat as one of the primary staple grains. Newly developed hybrid corn varieties yielded five to six tons per hectare when they were first introduced in Hebei province, compared to the three to four ton yields of the varieties they supplanted (Stone, 1990). By the late 1980s, hybrid corn covered 90 percent of China's corn planting area (Stone, 1990).

The development and rapid dissemination of high-yielding varieties reflected the success of China's breeding and extension programs. In the three decades prior to 1984, China's breeders were estimated to have developed over 3,000 cultivars of 41 economic plants (Tang, 1984). Research programs in China and their connections with institutions abroad, such as the International Rice Research Institute (IRRI), played a critical role in realizing yield expansion large enough to support China's growing population and industrializing economy (see Box 3.3).

Box 3.3 International Collaboration and China's Rice Breeding Programs: The Role of IRRI

A strong tradition of rice breeding both in China's research institutes and in the International Rice Research Institute (IRRI) in the Philippines made for a natural partnership in the development and dissemination of high-yielding rice varieties. Supported by the Ford and Rockefeller Foundations since it was established in 1960, IRRI has became one of the world's leading institutions for breeding improved rice varieties suited to growing conditions in tropical Asia. Although by the 1960s China's rice breeding programs were already well developed, IRRI scientists had virtually no contact with Chinese counterparts for political reasons. While a few varieties reached the mainland through unofficial channels, contact between IRRI and mainland China did not begin until almost ten years after IRRI was founded (Chandler, 1982).

Formal exchange between IRRI and China began when Philippine President Ferdinand Marcos presented seeds of the IRRI rice line IR20 to a Chinese trade delegation in 1972 (Chandler, 1982). In 1974, the first face-to-face meeting between IRRI and Chinese scientists marked the beginning of a long and mutually beneficial collaboration. The Chinese scientists shared their techniques for developing three-line hybrid rice, as well as research on natural fertilizers and the role of anther culture in rice improvement. Plant materials developed in China with beneficial properties such as early maturity, high seedling vigor, and high yield potential were also used in IRRI breeding programs (Chandler, 1982). Later, IRRI would play an important role in the development of blight-resistant rice in China.

Collaboration with IRRI greatly benefited China's rice farmers. Surveys of rice varieties planted in China estimated that around 20 percent were derived from materials developed at IRRI (Hu et al., 2000). These varieties included mostly members of the indica sub-species. In the south, where indicas are most widely grown, IRRI varieties are particularly common. In Hunan province (one of China's largest hybrid rice growing provinces, located directly south of the Yangzte River in central China), IRRI materials account for more than 40 percent of the planted area (Hu et al., 2000). In the north, where the japonica subspecies is more common, fewer IRRI varieties have been planted.

Collaboration has focused heavily on the exchange of varieties and techniques used in the development of hybrid rice varieties. Two of China's most successful hybrid varieties, *Shanyou* 63 and *Shanyou* 10, have parent materials from IRRI. In provinces that cultivate a higher than average percentage of hybrids, IRRI varieties account for a larger share of the crop. In turn, IRRI scientists learned hybrid rice breeding techniques from their Chinese counterparts (Chandler, 1982).

Fertilizer

The ability to realize yield increases in China without expanding cultivated area owed much to the introduction of nitrogen-based chemical fertilizers. In 1949, only two small fertilizer plants in China were producing ammonium sulfate. Orders were later placed in the early 1960s for five medium sized plants, which supplied 50 percent of China's nitrogen needs by 1965 (Smil, 2004). To further ensure that grain yields could keep pace with food demand, the Chinese government placed an order for 13 large, modern ammonium sulfate plants, immediately following United States

President Richard Nixon's visit to China in 1972 (Smil, 2004). Nitrogen-based chemical fertilizer use in China has increased more than fivefold between 1978 and 2005 (*China Statistical Yearbook,* 2006). China today remains heavily dependent on fertilizer, with Jiangsu, Zhejiang, Fujian, and Guangdong provinces in southeastern China together using more nitrogen-based chemical fertilizer than any other region in the world (Smil, 2004).

Irrigation

Water inputs also played an important role in realizing the yield potential of high-yield varieties. Irrigated area was expanded in the 1950s and 1960s, in part due to the revitalization of systems built in the 1930s. Though many of the systems newly built in the 1950s failed as a result of design flaws, the number of irrigation projects continued to increase throughout the 1970s. During the 1980s, in the midst of broader agricultural reforms, the expansion of irrigated area stagnated. Though the magnitude of this trend has been partially attributed to the exaggeration of statistics in prior years, its direction is not disputed. Furthermore, scholars have noted that the reductions in irrigated area during this period were particularly pronounced in relatively poor areas, many of which were bypassed by construction efforts (Stone, 1990).

Pesticides and Weather Protection

Pesticide use did not increase dramatically in China until the early 1980s, when organophosphate pesticides were first introduced from overseas. Though most of these compounds were banned in advanced industrialized countries during the 1970s due to their adverse human health and environmental effects, China continued to rely on some of these toxic pesticides up until the mid-1990s. The use of pesticides in China today will be discussed in more detail in later chapters in relation to the development of insect-resistant crops.

On many of China's marginal agricultural lands, the use of plastic sheeting has helped to guard crops against extreme temperatures and weather conditions. These polyvinyl chloride and polyethylene coverings were essential to supplying suitable conditions for high-yielding rice varieties to reach their maximum yield potential in some areas. Plastic sheeting was first used in 1958 and spread to many parts of the country (Stone, 1990). In 1979, the China Agricultural Inputs Corporation began to manufacture the covers on a large scale. Within two years, the sheeting covered 2.2 million hectares, permitting the cultivation of rice and corn crops in areas where it was previously difficult or impossible (Stone, 1990).

Challenges Ahead

China faces many of the same challenges that have been associated with the introduction of Green Revolution technologies around the world. Many of the productivity gains associated with early generations of high-yielding varieties and associated inputs have been reached. Since the research system has been a strong contributor to expanded agricultural productivity, a slowdown in public investment may hamper its ability to supply new varieties in the future.

Although responsible for great productivity gains, current irrigation practices in China are straining water supplies, resulting in conflict and promoting land degradation in some areas. Some lands have suffered so severely that crops are no longer viable there. As large cities continue to place demands on China's water supply, less will be available for agriculture, which accounts for around 70 percent of the country's freshwater consumption. Water constraints are not reflected in usage prices. If the energy costs of extraction are ignored, water is essentially free of charge in many areas (Huang, Rozelle, Howitt, Wang, & Huang, 2006). Water scarcity has also become a source of conflict. For instance, 1,000 villages in Shandong province on the central eastern coast fought for two days when police tried to prevent them from using makeshift irrigation schemes (Economy, 2005). Over-irrigation in some areas has left soils highly saline, which has in turn lowered the yield potential of today's available crop varieties. Especially in the arid north, drought and chronically low groundwater supplies have caused problems for local farmers. One proposed solution has been to pipe water from the south to the north, where groundwater deposits are only a fourth of the endowment in the south (Economy, 2005). However, this redistribution is costly and will not fix the underlying problem, which is that China is using its water resources at an unsustainable rate.

The rising application of fertilizers and pesticides has had severe effects on surrounding ecosystems. The chemical byproducts of heavy fertilizer inputs enter groundwater and cause changes in local ecosystems by altering the chemistry of soils and groundwater. Many of these changes are only reversible if nitrogen inputs are reduced, but that would lower yields, and even then the residual effects of heavy nitrogen applications would take years to reverse. Meanwhile, higher applications of nitrogen fertilizer are resulting in a declining yield response (Zhu & Chen, 2002). Fertilizer also accounts for the largest fraction of agricultural input costs in China, nearly 50 percent by some estimates (Yan et al., 2006). Chemical pesticide residues have added to environmental pollution and have taken a severe toll on farmer's health.

Rising average incomes in rural and urban areas have also led to significant changes in dietary choices, which are increasing the per capita demand on China's agricultural system. Protein-rich foods such as meat, milk, and eggs began to form a growing percentage of Chinese diets after the reforms, although staple grains, such as rice, corn, and wheat have remained the dominant source of calories in rural areas (Kueh, 1993). Demand for a broader range of vegetables and other cash crops has also grown in step with income. Since it takes a greater number of calories from grain to support livestock than the corresponding dairy and meat products provide, demand for grain is increasing faster than population growth. These trends may further complicate China's efforts to maintain self-sufficiency in grain in the future. Another related challenge involves improving the nutritional content of rural diets, as large segments of the rural population do not consume sufficient quantities of micronutrients, including iron, zinc, and Vitamin A (Hu, Tong, Oldenburg, & Feng, 2001; Zhang, Binns, & Lee, 2002).

Conversion of arable land to non-agricultural uses is further straining China's agricultural sector, although the impact of this trend on total agricultural land is disputed. Economic growth and corresponding increases in land prices have led to the loss of valuable crop land to development. One recent estimate indicates that China has lost one-fifth of its arable land to a combination of soil erosion and economic development since 1949 (CIA, 2006). When encroachment of urban areas threatened grain self-sufficiency in the mid-1990s, the government launched a land reclamation program in an effort to promote grain cultivation (Economy, 2005). However, this trend looks unlikely to continue, since government conservation programs are limiting further conversions, and land degradation and development pressures are unlikely to abate anytime soon.

The economic and social impact of the new agricultural technologies is hard to measure precisely because of the confounding influence of shifts in rural policy in China throughout the last half of the twentieth century. It is clear that high-yielding varieties in particular made an important contribution to agricultural productivity growth in China over the last several decades (Hu et al., 2000). It is plausible that farmers with slightly better access to technology, a better natural resource endowment, or a stronger financial position were most able to reap the benefits of new technologies. However, in the case of China, many studies show that on balance poor and rich areas benefited roughly equally from the introduction of new crop varieties, and some hybrid varieties had a disproportionately large effect in poorer areas (Zhong et al., 1995; Hu et al., 2000). In the case of new agricultural technologies, public support for research and technology transfer in China helped to make them available to many farmers before the reforms began. Nevertheless, in the course of ongoing reforms, how to provide equal access to superior varieties and their associated benefits remains a prominent concern for China's agricultural policymakers.

If not addressed, these environmental and resource challenges will place an increasing burden on the agricultural sector, environment, and quality of life in China. As this chapter has described, past gains have relied on advances in technology and associated inputs. Yet these gains cannot persist indefinitely, especially since increasing fertilizers and pesticides would come at a high environmental cost. Hu et al. (2000) have suggested that addressing the challenges of feeding a growing population on limited (and in some places badly degraded) arable land will require new technological solutions. Indeed, much of the work presently underway in China's laboratories is aimed at addressing these problems. The next two chapters describe the scientific fundamentals of biotechnology, and how China has incorporated these advances into a national strategy for addressing present and future agricultural challenges.

4

Agricultural Biotechnology: New Tools for Ancient Practices

While Green Revolution technologies were being adopted on farms around the world, scientists were working to understand the biochemical and molecular basis for life. Their insights gave rise to new techniques for making highly specific changes in the properties of plants in ways not possible with traditional breeding methods. This area of research is often referred to as "modern biotechnology." However, the field of biotechnology (which encompasses any application of living systems to develop new products and processes) is hardly new. Indeed, what we consider the emerging field of agricultural biotechnology is actually part of a continuum of breakthroughs in both science and agriculture that has unfolded since the earliest human civilizations.

Modern biotechnology has its origins in several important breakthroughs in the basic biological sciences. Building on Mendel's discovery that traits are passed from parents to offspring in the form of "factors" or genes, scientists of the early twentieth century connected this hereditary function with a long, negatively-charged molecule known as DNA. Found in every cell, DNA contains the genes that direct an organism's development. The discovery of genes and the mechanisms that enable their function laid the foundation for scientists to alter the activity of one or a few genes or transfer genes from one species to another in a highly specific manner. The term agricultural biotechnology refers to the application of these advances in the modern biological sciences to improve on products and processes in agriculture. One of these advances, genetic engineering, inspired considerable controversy when it was first applied to modify the genetic codes of crops to create "transgenic" varieties. This chapter describes the scientific advances that underpin this toolbox of techniques.

The Molecular Basis of Life

In the early twentieth century, many scientists pondered how the "factors" that Mendel had discovered could translate into the astonishing diversity of traits found in living organisms. The answer to this question would help not only to solve an age-old mystery, but to eliminate much of the guesswork from agriculture and medicine. Indeed, for years doctors prescribed drugs with little understanding of why they

V. J. Karplus and X. W. Deng, *Agricultural Biotechnology in China.*
© Springer 2008

were effective. Breeding programs took months or years to determine if particular crosses would yield the desired results. Scientists, mostly in Europe and the United States, therefore worked diligently to uncover the properties of the "factors" and reconcile existing knowledge with new discoveries.

A few critical pieces of the puzzle were already in place. In the 1600s, a crude microscope fashioned by the Dutch clothing dealer Antonie von Leeuwenhoek revealed that the world was covered by a ubiquitous and diverse population of single-celled organisms invisible to the human eye. He called these organisms "animalcules." This discovery led others to develop even more powerful lenses and observe that every living organism is made up of these microscopic units, which later became known as cells. Several hundred years later, in 1869, chemist Friedrich Miescher isolated a mildly acidic compound from the cells of several higher organisms. This compound is known today as deoxyribonucleic acid, or DNA.

In the early 1900s, geneticists showed that chromosomes—the long, thin threads of DNA contained in every cell—contained the blueprint for inherited traits by noting that the sorting of chromosomes during cell division tracked with eye color in flies. In 1944, scientists Oswald Avery, Colin McLeod, and Maclyn McCarty demonstrated conclusively that this DNA molecule was indeed responsible for trait inheritance by showing that transferring DNA from one type of bacteria to another caused the second to express characteristics of the first (Avery, McLeod, & McCarty, 1944). The structure of DNA was solved by James Watson and Francis Crick in 1953, who relied on data generated by Rosalind Franklin, an X-ray crystallographer, and Linus Pauling to show that the molecule formed a double helix, a structure comprised of two side-by-side strands held together by paired chemical groups.

DNA was later found to be capable of replicating itself, and in doing so, providing the blueprint for the synthesis (or "transcription") of messenger molecules (called messenger RNA). Strands of messenger RNA are then translated by still other molecules into proteins. Proteins, along with sugars, lipids, and other compounds, comprise the building blocks of a living organism. Think of the process—known as the "central dogma" of biology—as transcribing a message (say, original musical compositions) in mobile form (published music textbooks) and then translating it into the diversity of life's components (the harmony of sounds in a piano concerto). In the analogy, each note corresponds to a gene, and like the notes in Mozart's concertos, expression of a gene can be on or off, or strong or weak. Discoveries over the last several decades have revealed that this analogy is oversimplified. Since it was first proposed, the central dogma has been modified to account for cases where RNA can serve as a blueprint for the synthesis of DNA as well as for the various and interdependent roles that DNA, RNA, and proteins often play in regulating important steps in the overall process.

A DNA sequence is comprised of molecular units known as nucleotides linked together in linear fashion. Each nucleotide's identity is defined by one of four molecules (adenine, thiamine, cytosine, or guanine or A, T, C, and G, respectively) known as bases. Each base is connected to a five-carbon sugar, and then linked to adjacent nucleotides through a phosphate group. Genes correspond to segments of DNA defined by a sequence of hundreds or thousands of nucleotides. Only a small fraction of DNA sequences in higher organisms correspond to genes that supply the

Box 4.1 Key Terms in Biotechnology

DNA – A molecule comprised of a unique sequence of the four nucleotides, adenine (A), cytosine (C), guanine (G), and thymine (T), and encodes the blueprint for an organism's life processes.

Messenger RNA (mRNA) – A molecule generated by assembling the sequence complement of DNA, pairing T with A, A with U (uracil), G with C, and C with G, which is in turn decoded in the process of protein synthesis.

Nucleotide – A chemical compound, often a single unit of a DNA or RNA sequence, comprised of a base (A, T, C, G or U) connected to a five-carbon sugar and at least one phosphate group.

Protein – The components of molecular machines that are comprised of amino acids and underpin most functions in living cells.

Amino acids – The distinct chemical compounds that comprise the sequence of a protein. Each individual amino acid is specified by one or more sequences of three nucleotides in a strand of messenger RNA.

Agricultural biotechnology – A set of tools and techniques that are used to understand and alter the genetic makeup of crops, livestock, or microorganisms to increase productivity or add other desired traits.

Conventional Breeding – Visually assessing the qualities (phenotype) of offspring and selecting superior parents for reproduction to reinforce or introduce desired traits.

Molecular Breeding – Using knowledge of an organism's DNA sequence to select superior parents for breeding or to confirm traits in offspring.

Genome – All of the genetic information contained in an organism, generally in the form of a DNA sequence, which contains individual genes that influence traits and development.

Genomics – A new field dedicated to the sequencing of organism genomes and subsequent analysis to determine gene function and interaction among genes and gene products.

Tissue Culture – The process of artificially propagating plant tissue or organs, or reproducing a whole plant from a single or few cell(s).

Transgenic Crop – A crop containing genetic material that originated from a plant or another potentially unrelated organism and that is introduced using genetic engineering techniques.

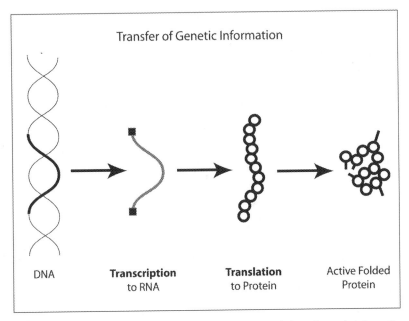

Fig. 4.1 The "central dogma" of biology that describes transmission of genetic information.

blueprints for proteins. These protein-encoding genes are located within long stretches of DNA, most of which corresponds to so-called "non-coding" regions. The gene is preceded by a promoter and ends with a stop signal, both of which are comprised of distinct nucleotide sequences. Protein complexes called RNA polymerases recognize the promoter and synthesize the gene's complementary strand by recruiting each base's corresponding partner (A with U or uracil, C with G, G with C, and T with A), with the help of many other proteins. This strand, known as messenger RNA or mRNA, is then translated by other molecular machinery into its corresponding protein sequence (see Figure 4.1). The complexity and adaptability of living organisms stems in large part from the vast number of possible protein sequences a single DNA molecule can encode, and the wide range of additional factors and processes that can modify them.

The Tools and Techniques of Agricultural Biotechnology

The foundation of agricultural biotechnology consists of laboratory tools and techniques developed since 1970, which can be broadly grouped into three categories: genomics, transgenic technology, and tissue culture. In practice these categories overlap significantly, and may even require techniques from other categories to succeed. For example, tissue culture techniques are typically used to regenerate a transgenic plant. Transgenic techniques have generated the most widespread controversy, but are one of many other developments in biotechnology that have numerous applications in agriculture, and especially in plant breeding.

DNA Sequencing and Genomics

With a basic understanding of trait expression in hand, scientists began to develop new techniques to elucidate the role of individual genes. In 1972, researchers discovered one category of proteins, known as restriction enzymes, capable of severing DNA at a particular site by recognizing the flanking sequences. By using these enzymes as molecular scissors, biologists could isolate the sequences they wanted to study.

Restriction enzyme technology provided researchers with a powerful tool to parse genomes into manageable pieces. The nucleotide sequences of these DNA fragments can then be read off one by one and reconstructed to reveal the sequence of entire genes. Improvements in sequencing technology have made it possible to determine the genome sequences of many organisms, including humans, mice, and rice. With sequence information in hand, breeders can better predict the outcomes of crosses or design drugs that counter specific processes in a disease pathway. Genome sequences can also lend insight into an organism's potential response to certain environments, aiding in the determination of optimal planting time or soil nutrient composition. The suite of advances made possible by DNA sequencing technology is commonly referred to as genomics.

The ability to identify the sequence of base pairs that comprise an organism's entire genetic code (or genome) has opened a new chapter in modern biology. Sequence information is the first step needed to connect individual gene sequences with the appearance of particular traits. Typically, this mapping is performed by identifying plants with unique instances of a particular trait or traits (mutants), then comparing its DNA sequence with that of an unchanged plant (wild type). Mutations can occur naturally as a result of random changes in the genome, or can be accelerated by exposing the plants to an externally-applied agent, such as ultraviolet light, that induces mutations. New genetic tools have enabled scientists to identify at a base pair level the molecular changes that give rise to a mutant phenotype.

Over the last ten years, genomics research has accelerated rapidly, and sequencing is complete or well underway for hundreds of organisms. A major milestone for plant genomics was the sequencing of model plant *Arabidopsis*. Its genetic similarity to a number of crop plants allows it to be used in a wide range of breeding efforts. Since then, the rice genome has been sequenced, and sequencing efforts for other crop plants, including corn and tomato, have been announced.

The individual functions and interactions of the genes can be determined using new tools that resolve these interactions at the single gene or base pair level. The use of a partial or whole-genome microarray enables scientists to quantify the extent of individual gene expression for the whole genome simultaneously. By exposing identical plants to different environmental conditions and then performing microarray analysis, their gene expression profiles can be compared. This analysis also enables scientists to identify the gene or genes that are involved in a particular pathway, such as the response to certain types of diseases. Since the number of genes in plants is large (*Arabidopsis* has up to 30,000 and rice may have 40,000 or more) and their interactions are highly complex, powerful computers are required to interpret the microarray results. This need has given rise to the new field of bioinformatics, or the interpretation of biological information using computing technology.

Fundamental research has also moved beyond genomics in recent years to focus on the structure and interactions among the wealth of proteins and other compounds that influence an organism's response to its environment. Since a single gene can encode many forms of a protein (for instance, by rearranging, adding, or omitting certain segments), the field of proteomics has emerged to describe this added layer of complexity. The field of metabolomics, in turn, is focused on elucidating the biochemical pathways and complex interactions among them that underpin essential biological functions.

Genomics, and to an increasing extent proteomics and metabolomics, are helping to improve the efficiency of plant breeding. Knowing the precise sequence of an organism's genome makes the outcomes of particular crosses more predictable. Often, a desirable trait is associated with one or several gene(s) or a segment of DNA that differs from inferior varieties by as little as a single base pair. By identifying parents that contain this small difference, breeders can be more certain that the offspring will contain the favorable trait as well. The process of using differences of one or a few base pairs in the genome to increase the certainty of breeding outcomes is known as marker assisted selection. As sequence databases and understanding of complex interactions become more complete, the efficiency and overall performance of conventional breeding methods will increase as well.

Whole Plants from Single Cells: Tissue Culture

Tissue culture is an older, but still essential, technique that is used to regenerate plants and in some cases also to improve agricultural productivity. Intuitively, we think of plants as being reproduced from seeds, but whole plants can also be regenerated from a small piece (such as a stalk or leaf) or even just single cells. This technique can be used to create thousands of plants with a genetically identical makeup. In conventional breeding programs, tissue culture is primarily used to eliminate viruses or other plant pathogens that may persist in seeds. These pathogens, which have become embedded in the cells of their host plants, can be eliminated by selecting an unaffected cell for regeneration. Tissue culture is also important in the regeneration of a transgenic crop from single or a few transformed cell(s) (see Figure 4.2).

Anther culture is another method for obtaining genetically pure plants. The male reproductive cells (pollen) located on the anther are collected and used to regenerate whole plants that contain only the genetic contribution of a single parent, or half of a complete set of chromosomes. Under special conditions, this single set of chromosomes can be forced to copy itself, producing a full set of chromosomes. The resulting offspring have high genetic purity. Anther culture is sometimes preferred by breeders because they do not have to conduct the extensive backcrossing needed to generate a genetically pure strain.

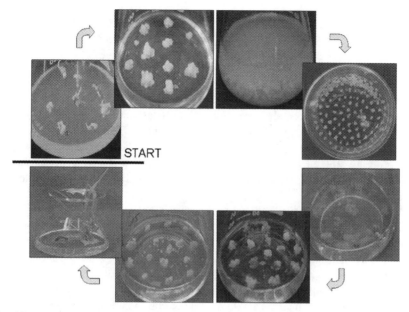

START

Fig. 4.2 A pictorial diagram of tissue culture of rice plants. The tissue culture process results in the generation of a whole plant from single rice cells. Photos courtesy of Dr. Xia Mian, Beijing Weiming Kaituo Agro-biotechnology Ltd., Beijing, China. (A color version of this figure appears between pages 72 and 73.)

Transgenic Technology

The same restriction enzyme technology which enabled scientists to isolate individual genes raised the possibility that genes could be transferred among different species. These types of transfers had been performed in early experiments with bacteria, using a process known as DNA recombination. This "cut-and-paste" application of the technology caught public attention from its earliest days of development in the early 1970s. The ability to transfer a genetic material from one species to another (a process also known as "transformation") was a significant step beyond traditional methods of sexual gene transfer, in which the genomes of both parents contribute genetic material to the offspring. It also enabled gene transfers to be verified before the plant had matured. Previously, scientists had to wait for a plant to reach maturity before results could be assessed. Recombination techniques are used routinely in laboratories today in molecular biology research. Genes can be introduced into model organisms (such as mice, worms, and model plants such as *Arabidopsis*), or some crops (such as corn, soybean, canola, tomato, and rice) to aid in identifying the physiological roles of the proteins they encode.

Plants were one of many organisms used in the first efforts to apply recombinant DNA techniques to create improved products. Part of the reason was that introducing novel genetic material into plant cells is generally easier than introducing it into animal cells. In plants, cells that can give rise to a whole transgenic plant are available in great abundance. Most plant cells can be considered as the "stem cells" that are required to regenerate a new organism. By contrast, in animals, most cells that assume a specific function in an organism "forget" how to initiate development, although all the necessary genes are still present in their cells. As a result, "stem cells" in animals are very limited in number.

Plant cells also differ from animal cells in other important ways. For example, plant cells contain chloroplasts, or small intracellular compartments that aid in harnessing energy from absorbed light. Chloroplasts are thought to have once been independent bacteria and, for that reason, contain DNA, which make them potential candidates for some genetic engineering applications. Plant cells also contain a cell wall, or a thick coating around the cell membrane that gives the cells a rigid structure. Animal cells possess only the flexible, more permeable cell membrane.

Many methods can be used to create a transgenic plant. The method preferred depends on whether a plant is classified as a monocot or dicot. Monocots develop a single cotyledon in their seed and can be identified by the parallel veins in their leaves. Common monocots include the world's grains—rice, wheat, rye, corn, oat, and sorghum, among others. Dicots, on the other hand, have two cotyledons and are characterized by branching veins in their leaves. Tobacco, cotton, soybean as well as most vegetables are dicots. This distinction is important because some methods of transforming plants will work efficiently for either monocots or dicots, but not both.

One common method of transferring DNA into plants involves infecting the plant with a soil bacterium called *Agrobacterium tumefaciens*. In the wild, the bacterium selectively transfers a gene that causes tumor growth in plants classified as dicots. In the laboratory, recombinant DNA techniques are used to replace the bacterium's tumor growth gene with a desirable gene of choice, such as one encoding disease resistance or herbicide tolerance. The bacterium will then deliver the desired gene instead of the harmful one into plant cells during infection, endowing the plant with the new trait. Another method for transferring foreign genetic material employs a "particle (or gene) gun" to bombard a cell with individual tiny gold particles coated with the genetic material to be transferred.

Once the gene has been taken up by the receiving cells, it is important to confirm that it is integrated and fully functional in the genome of the host plant. For this purpose, the introduced DNA sequence contains not only the sequence of the gene or genes to be introduced and a promoter (the nucleotide sequence that directs an RNA polymerase to start transcription), but also a marker gene. Marker genes are expressed only when the transformation is successful. Researchers distinguish transgenic varieties from non-transgenic varieties by enabling the latter to grow in the presence of certain compounds, such as antibiotics, which would otherwise kill the cells. Any cells that are able to survive and replicate on the medium must also be transgenic. Since many scientists and regulators are concerned about the use of antibiotic-resistant marker genes in large-scale agriculture, less controversial alternative strategies are now in use or being developed in many laboratories.

Successfully transformed plant cells are typically cultivated using standard tissue culture techniques as explained above. Scaling up seed production for use on farms requires crossing the transformed plant with local varieties, and is usually arranged by seed companies, which pay farmers or extension stations to grow crops from which the seeds are later collected and sold.

Applying Agricultural Biotechnology in Crop Design

With these new tools and techniques, scientists have the potential to alter the properties of crops more rapidly and precisely than ever before. For instance, using microarray or other detection tools, breeders could quickly check for the presence or expression of a particular form of gene or genes. Also, the ability to transfer genes one at a time eliminates the possibility that disadvantageous genes will be reinforced along with advantageous genes during backcrossing. With the ability to identify genes and better predict interactions among them, genomics helps breeders to design crosses that minimize the resulting weaknesses. The ability to selectively transfer genes avoids this risk altogether.

Transgenic technology opens up a broad range of new possibilities for scientists and breeders. So far the technology has focused on traits that confer benefits to farmers, as did many of the Green Revolution advances. For example, genes that confer herbicide tolerance or insect resistance help to reduce yield losses, enabling farmers to reap a greater harvest with reduced need for expensive and environmentally taxing agricultural inputs. Currently, consumers have gained few benefits from the technology. Yet developers claim that the technology could eventually be used to create foods that are more nutritious, flavorful, or even address global public health problems by providing easily deliverable vaccines.

Biosafety Concerns

It is important to understand how non-transgenic and transgenic applications of agricultural biotechnology can be used in the creation of new crop varieties and the reasons why each application has raised its own set of concerns. Many applications of agricultural biotechnology do not rely on gene transfers among organisms. Instead, they include procedures such as genome sequencing and the mapping of a gene to its functional role in a particular pathway. In plants, selection of the genetically robust for breeding has been a mainstay of agriculture since its origins, and therefore non-transgenic applications of biotechnology to agriculture have not raised serious concern in their own right.

In contrast, scientists have not always recognized as safe the recombinant DNA techniques used to develop transgenic organisms. When the technology was first demonstrated in the early 1970s, communities living in close proximity to recombinant DNA laboratories expressed many concerns, among them that misuse or accidents might produce toxic or disease-causing organisms. Even within the scientific community, there was widespread unease about how to proceed. After meeting several times to discuss the potential risks, many of the first scientists involved called for a voluntary moratorium on recombinant DNA research in 1974. In 1975, scientists reconvened at the Asilomar Conference Center near Monterey, California, to develop guidelines for recombinant DNA research (Berg, Baltimore, Brenner, Roblin, & Singer, 1975; Berg & Singer, 1995). By engaging in voluntary regulation, early developers helped to ease public concerns, while allowing development of the new technology to proceed.

Several properties of transgenic techniques have prompted public concern. Given that genes can be transferred into plants from unrelated species as well as more closely related ones, many worry that in its new environment an introduced gene could have unpredictable and unintended effects. For example, if a peanut gene is introduced into soybean plants to improve the nutritional content of soy oil, can consumers allergic to peanuts safely consume the transgenic soy product? Others worry that interactions between the novel gene's encoded protein and proteins native to the host organism might result in the production of toxic compounds. Since every interaction of a novel gene's protein product cannot be predicted with certainty, such scenarios are hard to rule out entirely. Identifying allergenic or toxic properties of new plants or food products is already a routine part of food safety systems in many countries, but the question has remained whether and how these rules need to be revised or expanded to cover transgenic crops. These issues will be discussed in more depth in Chapter Eight, along with discussion of the biosafety concerns raised by transgenic crops in the field.

Global Diffusion of Agricultural Biotechnology

Although initial understanding of gene function and recombinant DNA technology emerged from laboratories in Europe and the United States, the tools to apply these breakthroughs to agriculture have since spread widely, and research laboratories with capacity to apply them have been established on nearly every continent. In many parts of the world where Green Revolution technology has taken root, laboratories have adopted basic genomic tools to strengthen local breeding efforts. Other countries have been less keen or able to develop agricultural biotechnology techniques for a variety of reasons—research costs, lack of trained personnel, and public hesitation or rejection of transgenic applications. Still, many countries in both the developed and developing world have shown interest in applying this set of tools and techniques to address agricultural challenges.

5

Agricultural Biotechnology Takes Root in China

In the early 1980s, agricultural biotechnology ranked among the most prominent components of a broad national initiative to invigorate China's science and technology. The leaders of China's reform and opening program had only to look to the industrialization of Western Europe to note the contribution of technological progress to economic growth and global influence over the past several centuries. The "space race" between the United States and the Soviet Union, the postwar Japanese economic miracle, and recovery in Western Europe provided more recent, tangible examples. A long history of inventiveness and reverence for scholarship, combined with an emphasis on science and technology in Marxist-Leninist doctrine, created a favorable climate for increasing investment in research programs during the reform years. In the early 1980s, China's new leader, Deng Xiaoping, announced that China would employ both foreign expertise and domestic policy changes to reinvigorate science and technology on the mainland.

A far-reaching effort to modernize the nation's science and technology could not ignore its largest sector, agriculture. Despite efforts to promote the development of industrial sectors prior to the beginning of the post-1978 reforms, the agricultural sector was still China's largest employer and main contributor to national income when the reforms began. A large and growing population and scarcity of arable land added to the urgency of research pursuits. Throughout the 1960s and 1970s, the success of research efforts in developing high-yield varieties that were widely planted on farms provided a strong case for redoubling investment in agricultural science and technology. By the 1980s, early successes abroad suggested that applications of biotechnology might produce the next generation of yield-enhancing technologies.

When China's investment in agricultural biotechnology began in the early 1980s, it was driven mainly by two emerging national priorities: building a modern scientific enterprise and meeting the challenges facing the agricultural sector. In this chapter, we first recall the early roots of plant biology and biotechnology in China. We then trace the origins of agricultural biotechnology programs during the reform period.

V. J. Karplus and X. W. Deng, *Agricultural Biotechnology in China.*
© Springer 2008

Origins of Modern Plant Biology in China

Basic research in modern plant biology first reached China in the 1930s and 1940s, principally by way of Chinese students returning from overseas training. In the 1930s, Drs. Pei-sung Tang and Tsung-lee Lo set up China's first training programs in plant biology and published work that spanned several plant biology fields (Tang, 1981; Chen, Cheng, Cheng, & Tang, 1945; Tang & Wu, 1957; Tang & Loo, 1940). Around the same time, Dr. Hung Chang-Yin made several discoveries on mechanisms of photosynthesis (the process by which plants convert sunlight into a usable form of energy) (Yin & Sun, 1947; Yin & Tung, 1948). Chinese plant physiologist Dr. Lou Chenghou studied together with Nobel Peace Prize Laureate Dr. Norman Borlaug at the University of Minnesota before returning to China to begin a long research career at China Agricultural University. Lou and a handful of his contemporaries went on to make important contributions to their fields and published several articles in international journals (Lou, 1945; Hsueh & Lou, 1947). This spike in early citations was followed by a precipitous drop after 1949. With the exception of limited exchange with the Soviet Union, China's doors to overseas collaboration remained closed until the late 1970s.

Science and Technology in the Pre-Reform Period: 1949–1978

Although significant investment in biotechnology and other "high technology" areas did not begin until the mid-1980s, interest in developing science and technology was part of national agendas from the earliest years of the People's Republic. The research system described here (with the Chinese Academy of Sciences at its core) is distinct from the agricultural research system under the Chinese Academy of Agricultural Sciences (CAAS) described in Chapter Three. The Chinese Academy of Sciences (CAS) was set up to focus on fundamental (also referred to here as "basic") research, which refers to an inquiry into the properties of the natural world through systematic, creative experimentation. By contrast, applied research is usually aimed at developing or improving a specific process or product. The CAS was established in 1949 by combining the former Republic's Central Academy of Sciences and the Beijing Academy of Sciences (Hamer & Kung, 1989). The CAS was initially subordinate to the Ministry of Education, then later elevated to ministry status in 1954 (Tang, 1984). The CAS institutes covered a broad range of scientific disciplines, including basic research in physics, chemistry, and biology.

The organization of scientific research in China was modeled after its counterpart in the Soviet Union, whose leaders in turn provided the new government of the People's Republic with funds, technology, and advice. This direct influence was severely weakened when Soviet leader Nikita Khrushchev cancelled all 257 Sino-Soviet scientific cooperation programs, revoked construction contracts, and recalled over 1,000 technical advisors in the late 1950s (Tang, 1984). Nevertheless, early Soviet involvement left behind a legacy that heavily influenced the subsequent evolution of China's scientific institutions and policies. The tradition of political leadership of the national science and technology enterprise was firmly established, both as a result of Chinese historical precedent and Soviet influence. This leadership manifested itself in a series of initiatives to promote research activities. The first central policies on science and technology were embodied in a twelve-year plan (1956-1967) to support research on atomic energy, jet and rocket engines, electronics, and other areas. After announcing the original plan had been completed five years ahead of schedule, a new ten-year science plan was drafted in 1963 (Tang, 1984). This pattern of announcing research priorities and targets spanning fixed time frames (usually five, ten, or fifteen years, based on the Soviet model of the Five-Year Plan) has been a mainstay since the People's Republic was founded.

The weak linkages between fundamental and applied research remained a serious shortcoming of the Chinese system. Although CAS was intended to be the nation's premier source of cutting edge technology, it remained relatively isolated from state-owned production units, which answered to their respective industrial or agricultural ministries. As a result, breakthroughs within CAS diffused only slowly out of the CAS system, and were rarely applied in industry. Engineering and science disciplines were taught in separate institutions, further distancing fundamental and applied activities (Tang, 1984). Although fundamental research in plant biology was housed within the CAS, agricultural research was mostly conducted in the CAAS, with limited mobility between them.

The Soviet-style organization of China's research system after 1949 was strongly reflected in the nation's scientific successes and shortfalls. Aside from an aggressive military research program (which included development of the atomic bomb), most research proceeded without attention to national or industrial needs. Since full employment was an explicit goal of the communist system, research institutes continued to support nonperforming personnel and research programs. Over the 1980s and 1990s, observers recognized that the system was highly redundant, inefficient, and insulated from the needs and interests of society (Suttmeier, Cao, & Simon, 2006; Huang, Hu, & Rozelle, 2002b).

China's Agricultural Biotechnology Investment

Although agricultural research made important contributions to crop yields during the pre-reform period, overall China's science and technology research in the late 1970s remained far behind the international frontier. For several decades, scientists on the mainland had remained isolated from breakthroughs abroad, and particularly from the emerging field of molecular biology and biotechnology. By the end of the 1970s, recombinant DNA experiments had become routine in the United States and Europe. In China, however, most students were being educated in schools that did not provide even basic training in biochemistry and molecular biology. Meanwhile, Chinese publications in international journals had slowed to a stop in nearly every field.

National policies to strengthen scientific competency and agricultural production helped put agricultural biotechnology on the national science and technology agenda as early as 1978 (Hamer & Kung, 1989; Wu, 1983). Optimism abroad that recombinant DNA and other molecular biology techniques might offer major break-throughs in agriculture and medicine attracted the attention of China's top leaders, many of whom had been trained as scientists or engineers themselves. Biotechnology's relative youth added to its appeal, as some advisors believed that Chinese researchers could close the gap with developed countries and move ahead with seminal discoveries. Other proven and potential advantages of China's economic situation—a highly literate population, inexpensive manufacturing capacity, and growing role as global exporter—suggested that the industry might grow internationally competitive within a few decades.

During the 1980s, the government announced several funding programs to support biotechnology research. The most prominent initiative, the National High Technology Research and Development Program (or 863 Program), was conceived at the recommendation of four respected Chinese scientists, Wang Daheng, Wang Ganchang, Yang Jiachi, and Chen Fangyun. Named for its launch in March of 1986, the 863 Program was intended to strengthen both fundamental and applied research to address national needs (See Box 5.1). New structures and procedures were created to administer program funds and identify long-term priorities. The original priority research areas included biotechnology, information technology, energy resources, material science, automation technology, marine science, and space/laser technology. Of these areas, biotechnology was designated top priority (Chen & Qu, 1997).

The launch of the 863 Program contributed to a dramatic increase in funds for science and technology development. Prior to 1985, China's investment in agricultural biotechnology was very small. Over a period of 15 years (1986 to 2000), 10 billion yuan (equivalent to US$3 billion in 1985 dollars) was allocated to science and technology development under the 863 Program (Huang & Wang, 2002). Of the 1.3 billion yuan allocated to biotechnology from 1986 to 2000, half was allocated to agricultural applications (Huang & Wang, 2002). This sum was supplemented by other funding programs that included agricultural biotechnology in their scope of support. International funding agencies, in particular the Rockefeller Foundation and World Bank, also provided support for China's agricultural biotechnology development (Normile, 1999; Tang, 1984).

Other early programs to support agricultural biotechnology were incorporated into China's Seventh Five-Year Plan (1986 to 1990). Initially, funding for biotechnology research spanned many sub-fields, including plant genetic engineering, cell engineering, enzyme engineering, and others (Hamer & Kung, 1989). Responsibility for administering research programs was spread among the CAS, State Education Commission, and the Ministries of Agriculture, Health, Medicine, and Light Industry (Hamer & Kung, 1989). Grant classes varied by size and duration, and included General Awards, Key Projects, Frontiers of High Technology Projects, and Young Scientist Awards. The final category was reserved for scholars under 35, and aimed at attracting promising young scholars to return from studies overseas.

One common theme of China's early biotechnology funding programs was its focus on applied research. With the exception of the National Natural Science Foundation of China (NNSFC), which was founded in 1986 (see Chapter Nine for more information), China's emerging research programs focused mostly on the research areas expected to yield the most rapid and profitable results. Many leaders viewed this emphasis as necessary to modernize the country. As described in Chapter Three, core funding for institutes engaged in applied research was reduced to encourage the development of marketable products. Research institutes adapted to the changes by reinventing themselves as companies, by performing contracted research for industry, or by transferring technology to outside firms.

Since 1990, many of these early programs have expanded in scope and new programs have been added. In 1986, funding for agricultural biotechnology barely exceeded US$4 million. Yet by 1995, funding had grown to US$10.5 million, and by 2000 had reached US$38.9 million (Huang & Wang, 2002). In addition to continued strong support under the 863 Program, additional agricultural biotechnology funding initiatives were launched, including the National Basic Research (973) Program, Special Foundation for Transgenic Plants, and the Key Science Engineering Program. Established in 1997 as a follow up to the 863 Program but with greater emphasis on fundamental research, the 973 Program includes a focus on plant biology projects with biotechnology applications. In 1999, the Special Foundation for Transgenic Plants was set up to focus specifically on the development and commercialization of transgenic crop plants. The Key Science and Engineering Program, which supports a wide variety of research areas, allocated 140 million RMB (US$17 million) in 2000 to one particularly large project to improve plant material quality (Huang & Wang, 2002). Since then, continued support for agricultural biotechnology research has been evident in the redoubled funding commitments in the Tenth Five-Year Plan (2001 to 2005) and Eleventh (2006 to 2010) Five-Year Plan, which will be discussed later in Chapter Nine.

Administration of Research Agendas

China's top decision-making authorities assumed prominent roles in defining early research agendas. In its role as the government's top long term management and advisory body, the State Planning Commission (later the State Development Planning Commission or SDPC) formulated five-year and long-term plans for biotechnology development. Since it retains final approval on all ministerial budgets, the SDPC heavily influenced funding allocations for biotechnology until it was disbanded in 2003. The National Development and Reform Commission (NDRC) then replaced the SDPC and several other offices. Both the SDPC and later the NDRC have been highly influential in determining how control over funds for agricultural biotechnology is delegated to ministries and other initiatives (such as the 863 Program).

From the start, the State Science and Technology Commission (now the Ministry of Science and Technology or MOST) has played a prominent role in the funding and oversight of China's agricultural biotechnology research. Prior to 2006, agricultural biotechnology funding programs, including the 863 and 973 Programs, Key Science and Engineering Program, and Special Foundation for Transgenic Plants, were administered by the China National Center for Biotechnology Development

Box 5.1 The 863 Program and the China National Center for Biotechnology Development (CNCBD)

China's modern biology and biotechnology programs owe much to the early efforts of Chinese scientists working on the mainland and overseas. On January 10, 1983, Dr. Ray Wu, a Chinese professor of molecular biology based at Cornell University, wrote to State Councilor Fang Yi to emphasize the benefits of and opportunities for investing in biosciences education and research. In his letter, he recommended the formation of a national committee to coordinate research activities in genetic engineering. The following year, eight scientists from the United States and 20 from the mainland were appointed as advisors of the new China National Center for Biotechnology Development (*China Daily*, 1984). In addition to providing guidance on research and training activities, CNCBD's original mandate included the construction of major research facilities in the cities of Beijing, Shanghai, and Jiangmen (in southern China's Guangdong province). Plans for a Beijing center were later abandoned, and the center in Jiangmen suffered from partial withdrawal of the original support (Hamer & Kung, 1989). However, the CNCBD played a prominent role in setting China's early research priorities in biotechnology, both under the Seventh Five-Year Plan (1986–1990) and the newly-established 863 Program.

Named for its launch in March of 1986, the National High Technology Research and Development Program (863 Program) has remained one of the major domestic sources of funding for agricultural biotechnology research over the last 20 years. First announced by the Leading Group on Science and Technology under the State Council, China's highest government body, the 863 program included support for research in eight key areas. Among them, biotechnology was explicitly given top priority. The program's strong support for agricultural biotechnology echoes Deng Xiaoping's statement: "The ultimate outlet for agriculture in the future lies in bioengineering and sophisticated technology" (CNCBD, 2003).

At the outset, the 836 Program's biotechnology research budget (approximately US$3 million per year) was divided among agricultural (40 percent), medical (40 percent), and protein engineering (20 percent) projects (Hamer & Kung, 1989). Although nominally allocated according to a peer review process, successful grants are typically pre-selected through a complex negotiation process involving government administrators, CNCBD representatives, and the directors of the 863 Program. When the program started in 1987, one out of every five applicants was awarded a grant, which provided recipient laboratories with 500,000 to two million yuan over three to five years, among the most generous sums available at the time (Hamer & Kung, 1989). Today, the 863 Program supports many of China's National Key Laboratories and new institutions, such as the National Institute of Biological Sciences, Beijing (see Chapter Nine).

In addition to advising the CNCBD and 863 Program's biotechnology grants, Dr. Wu also initiated the China-United States Biochemistry and Molecular Biology Examination and Administration (CUSBEA) Program, which sponsored Chinese students in the biological sciences for doctoral work in the United States from 1982 to 1989. Although around 90 percent of the graduates took positions outside of China upon completion of their studies, the program helped to create a strong cohort of accomplished scientists that have been instrumental in shaping China's basic and applied biosciences research. For instance, many of these scientists have been invited to review grants for the National Natural Science Foundation of China. Some of the scientists trained under the CUSBEA Program have returned to take leadership or principal investigator positions in new institutes in mainland China as well.

(CNCBD), an office of MOST. Responsibility for agricultural biotechnology research programs has since been delegated to the China Rural Technology Development Center, also under MOST, primarily to integrate agricultural biotechnology with agricultural research agendas. The CNCBD retained administrative responsibility for biomedical and other fundamental biotechnology research programs, some of which support research with potential agricultural applications (see Box 5.1). Until recently, the Ministry of Agriculture (MOA) has held official responsibility for defining the agricultural research agenda in line with top national policy priorities, and MOST has retained discretion to translate this agenda into decisions to fund specific laboratories or projects. This situation has begun to change in recent years as oversight of some grant programs has been redistributed to give the MOA more responsibility, as will be discussed in more detail in Chapter Nine.

Scientists at universities and institutes have been the main beneficiaries of China's biotechnology funding programs. Many grants have been awarded to scientists at National Key Laboratories set up under the 863 Program. Housed mostly at existing universities or research institutes, 30 National Key Laboratories equipped to carry out biotechnology work were established during the 1980s and 1990s (see Figure 5.4). Of these laboratories, 15 had projects that focused on biotechnology applications in agriculture (mostly crops and livestock) (Huang & Wang, 2002). The national office of MOST also interacts with its offices at the provincial level to support regional and local initiatives as well.

New funding initiatives helped to increase the diversity and decentralization of China's biotechnology research enterprise compared with the pre-reform years. Prior to the reforms, nearly all research funding was allocated to institutes, principally at CAAS and CAS, and distributed according to targets set by the State Planning Commission. By establishing a competitive grant application process, the new funding mechanisms enabled universities as well as other top performers to gain a greater share of the funds. By the end of the 1980s, roughly half of funds were being disbursed to laboratories at universities (Hamer & Kung, 1989).

China's First Transgenic Crops

Within a few years after the launch of major funding initiatives, several laboratories in China began to develop agricultural biotechnology programs. Many of these programs included an emphasis on transgenic crop development. Among the first laboratories to develop a transgenic crop was Dr. Chen Zhangliang's group at Peking University. Trained in plant genetic engineering at Washington University in St. Louis, Missouri, and in labs supported by the Monsanto Company, Dr. Chen returned to China after completing his doctoral work in 1987. As a young scientist educated abroad in an emerging field (see Box 5.3), Dr. Chen attracted the support of the Chinese government as well as the Rockefeller Foundation. Within a few years, his lab had produced some of China's first transgenic crops, including virus-resistant tobacco and a virus-resistant tomato (Chen & Qu, 1997; Chen et al., 1992). The virus-resistant tobacco was field tested at agricultural stations in Liaoning province in the northeast. Scientists in Dr. Fang Rongxiang's laboratory at the Institute of Microbiology at CAS also developed virus-resistant tobacco, and conducted field tests in Henan province in the north central part of the country (Chen & Qu, 1997). Virus-resistant tobacco was made available to farmers in 1992, at which point it became one of the world's first transgenic crops to be grown commercially (Macilwain, 2003; Teng, 2004). At the Institute of Microbiology at CAS, Dr. Tian Bo's group developed virus-resistant potato plants, while Mang Keqiang's group developed insect-resistant transgenic corn (Chen & Qu, 1997).

The Biotechnology Research Center (now the Biotechnology Research Institute, see Figure 5.1) at the CAAS has also played a prominent role in early efforts to develop transgenic crops in China. Dr. Fan Yunliu initiated several programs focused on the development of transgenic insect-resistant crops after she transferred from

Fig. 5.1 Biotechnology Research Institute, Chinese Academy of Agricultural Sciences. Photo by authors. (A color version of this figure appears between pages 72 and 73.)

Fig. 5.2 Insect-resistant Bt cotton (right) and a non-transgenic control (left). Photos courtesy of Guo Sandui, Chinese Academy of Agricultural Sciences. (A color version of this figure appears between pages 72 and 73.)

the CAS Institute of Microbiology to CAAS in the mid-1980s. The insect-resistance gene developed in Dr. Fan's laboratory contained a gene from the bacterium *Bacillus thuringiensis* (Bt) that produces a protein that is toxic to the cotton bollworm and related insects. With funding from the Rockefeller Foundation's International Program on Rice Biotechnology, Dr. Fan began work on insect-resistant rice,

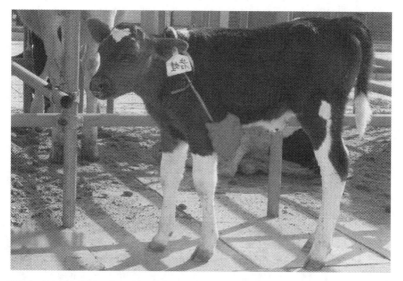

Fig. 5.3 A cloned calf developed by the State Key Laboratory for Agro-biotechnology at China Agricultural University. Photo courtesy of Dr. Chen Zhangliang. (A color version of this figure appears between pages 72 and 73.)

Fig. 5.4 National Key Laboratory for Agro-biotechnology, China Agricultural University. Photo by authors. (A color version of this figure appears between pages 72 and 73.)

but success was limited to model rice varieties not commonly planted on farms. In the early 1990s, Dr. Fan's student, Guo Sandui, successfully introduced the same form of the Bt gene construct into cotton plants, and had developed several transgenic Bt cotton varieties by 1993 (see Box 5.4 for more information on these early CAAS programs) (Pray, Ma, Huang, & Qiao, 2001). Later, CAAS researchers developed insect-resistant cotton varieties, with resistance conferred by either the Bt or Cowpea Trypsin Inhibitor (CpTI) gene (see Figure 5.2). By endowing cotton plants with insect resistance, several major pests could no longer survive on the cotton crop, drastically reducing the need to spray pesticides. To transfer the genetic material, developers employed both *Agrobacterium*-mediated transfer as well as a new method of pollen transfer, known as the pollen tube method, pioneered by Dr. Zhou Guangyu at the Shanghai Institute of Plant Physiology and others. Some scientists have reported that this method is not consistently reproducible (Hamer & Kung, 1989).

Box 5.2 From Revolutionary to Agro-biotechnologist and Plant Geneticist: Dr. Zhang Qifa, Huazhong Agricultural University

Several years after the founding of the People's Republic of China, the wife of a tractor driver in a rural village in China's central Hubei province gave birth to her firstborn son. As a boy, Zhang Qifa grew up working in the field.

During the harshest years of famine in 1960 and 1961, he recalls scavenging for a diet of wild grasses. Zhang attended a local brigade school for most of his youth, but during harvests the school would close and Zhang would tend to the fields. He later attended two years of middle school in nearby Jinzhou City, but when the Cultural Revolution began in 1966, he returned to the fields again. He witnessed criticisms of his high school teachers and long speeches by classmates who had joined the Red Guard, the communist youth organization. Later that same year, Zhang and several friends boarded a *jiefang* (liberation) brand truck for Wuhan, where they caught a train to Beijing to join masses of communist youth on a nationwide pilgrimage to see Chairman Mao in person. This journey marked the first time Zhang had ever traveled more than 20 kilometers from home.

Shortly after Zhang returned, Jinzhou devolved into chaos. A clash between the local army unit and party representation resulted in an all-out confrontation between the "pro-imperialist" and "revolutionary" camps. Zhang left for home in 1968 to help his family on the farm, joining members of work units that had been disbanded to the countryside. A year later he joined the brigade school, where he taught fourth grade until 1973, earning the party's endorsement as a worker-peasant-soldier-student, and was allowed to attend college (at that time, all students were admitted by party recommendation on the basis of behavior, not academic credentials). In 1973 he entered the new countryside campus of Huazhong Agricultural University, where he alternated his days between classes and field work for three years. After graduating with a degree in agronomy in 1976, Zhang took a job as a teaching assistant and taught wheat genetics and breeding. Amid the chaos and destruction surrounding him and the challenges of everyday farm life, he recalls feeling vaguely that something was missing. Realizing that many articles published in overseas journals on wheat breeding might offer insights into some of the major agricultural challenges his community was facing, he began to study English.

In 1978, two years after the Cultural Revolution had ended, Huazhong Agricultural University moved back to Wuhan, abandoning its countryside campus. Across China, classrooms were abuzz with news that students were being permitted to study abroad, and Zhang and his classmates signed up for the qualifying examinations. Though Zhang's self-study in English and other subjects helped him to pass handily, he lacked the connections needed to circumvent the constantly changing requirements for government permission to leave the country. In 1980, he was offered a position as a visiting scholar at University of California at Davis. Though Zhang wanted to be a student, not an employee, he refused to pass up a previously impossible opportunity to travel outside China's borders.

Though he could not officially take classes, Zhang was able to audit the typical graduate Ph.D. sequence, and proved an exemplary student. His advisor and department facilitated his switch into the doctoral track. He spent four and half years completing his Ph.D. in genetics before returning to China in 1987. Out of a sense of obligation as well as personal ties, Zhang returned to Huazhong Agricultural University with his wife, who had also left China to study veterinary medicine at the University of Edinburgh in Scotland. After spending a year traveling frequently to secure grant money, Zhang founded the College of Life Sciences and Technology with a combination of funds from his university, the Ministry of Agriculture's earliest biotechnology grants, and the National Natural Science Foundation of China. He started his lab with two centrifuges from Japan and 40 square meters of lab space, which has since grown to fill several buildings and is known as the State Key Laboratory for Crop Genetics and Breeding.

Today, Zhang Qifa integrates basic research on how plant genes work with his efforts to develop both transgenic and non-transgenic varieties that can better tolerate external stresses or are more nutritious. Behind his lab at Huazhong Agricultural University stretch demonstration plots, where Zhang encourages his students to learn about how plants respond in field conditions, while gaining proficiency in molecular biology techniques in the laboratory. His group has mapped over 20 important genes in the rice genome, identified corresponding molecular markers, and elucidated how several genes interact to produce complex traits. One particularly important advance described the genes that underpin photoperiod-sensitive male sterility, which is important in the "two-line" hybrid breeding system (Zhang et al., 1994). In addition to publishing many influential papers in China and abroad, Zhang's research has

Fig. 5.5 Rice strain engineered for resistance to bacterial blight (left) and its original (non-transgenic) strain (right). Photo courtesy of Dr. Zhang Qifa, Huazhong Agricultural University. (A color version of this figure appears between pages 72 and 73.)

focused on some of China's most persistent challenges to agricultural development, and has produced both transgenic and non-transgenic varieties that are resistant to bacterial blight, tolerant to drought or saline soils, and resistant to pests and diseases (see Figure 5.5). In 1999, he was one of the youngest members to be elected to the Chinese Academy of Sciences.

Zhang has also been one of biotechnology's most vocal proponents. His laboratory has developed a variety of crops using transgenic and non-transgenic techniques, including disease-resistant, drought- or salt-tolerant, or nutritionally-fortified crops. Following the controversy over transgenic rice (to be discussed in Chapter Eight), Zhang and fellow members of the Chinese Academy of Sciences submitted a letter to the biosafety regulatory authorities underscoring that several transgenic rice varieties had passed multiple safety evaluations, and that the crops should be recommended for commercialization.

Box 5.3 Transplanting Transgenic Technology: Dr. Chen Zhangliang, Peking University and China Agricultural University

Born in the early 1960s on a farm in rural Fujian province, Chen Zhangliang recalls how his father, a fisherman and rice farmer, would often return from work feeling ill from spraying pesticides. During the final years of the Cultural Revolution, Chen was assigned to the South China Agricultural University on the island province of Hainan, where he studied rubber processing, pathology and engineering. Towards the end of his studies, he acquired copies of the international scientific journals *Science* and *Nature*—a rarity in China at the time—where he came across several articles by Dr. Mary-Dell Chilton at Washington University in St. Louis. These articles described a series of breakthroughs on gene transfer mediated by *Agrobacterium tumefaciens*. Though he lacked formal training in genetic engineering techniques, he figured that based on her publication record, she must be highly accomplished in her field, and wrote to her to ask if he could join her lab.

Dr. Chilton encouraged him to apply, and helped to facilitate his admissions interview in nearby Guangdong province. Chen completed all of the application requirements, and in spite of his lack of experience, was accepted into one of the most advanced genetic engineering laboratories in the world. As a member of Dr. Chilton's lab, Chen participated in the first efforts to develop transgenic plants in 1983. When his mentor left the university later that year, Chen transferred to a laboratory run by Dr. Roger Beachy and supported by the Monsanto Company. In 1985, Chen helped to demonstrate the successful expression of a gene encoding a soybean seed protein in the seeds of a petunia plant (Beachy et al., 1985). After publishing several other papers on related topics, he defended his thesis and graduated in just three and a half years, a remarkably short time for a biology graduate program.

During his time in the United States, Chen participated in the development of some of the first transgenic plants at a time when scientists were beginning to understand their potential applications. He also assisted with the first field tests for one of the world's earliest transgenic plants, a virus-resistant tomato.

In 1987, at the age of 26, Chen received a call from the then Chinese Ambassador Han Xu in Washington, D.C. Ambassador Han had read about Chen's publications and speaking engagements in the *People's Daily*, a major Chinese newspaper. Since the 863 Program had just been launched, China's Ministry of Science and Technology was beginning to seek promising scientists from overseas to help rebuild China's research system. Chen could choose to start work in Beijing or Shanghai, and was given a fully funded round trip home to explore his options further. While in Beijing, Chen talked with the Vice Chairman of the National People's Congress at the time, who assured him that China was changing, and that democracy was beginning to take root. Drawn by the opportunity to lead in the modernization of China's scientific programs in a time of reform and opening to the world, Chen chose to return.

Chen did not consider his return to China risky. He reassured himself that he could always get a postdoctoral position in the United States, but felt compelled to "take a chance." Peking University awarded Chen a lab and an Associate Professorship, and he quickly initiated research on a variety of transgenic plants. His laboratory, the National Laboratory of Protein Engineering and Plant Genetic Engineering, was set up in 1986 with five or six people, but grew to over fifty members within a few years. Supported by the Rockefeller Foundation, the main focus of his lab's research was rice. Scientists in his lab studied rice antifungal properties and developed a rice transformation system (Chen & Qu, 1997). Due to the difficulties involved in engineering rice (a monocot), the lab expanded its focus to include transgenic tobacco and tomatoes as well. His lab also developed transgenic color-altered petunias and a virus-resistant sweet pepper, both of which were approved for commercial planting in the late 1990s (see Figure 5.6).

Chen's lab at Peking University field tested transgenic virus-resistant tobacco in the late 1980s, first in Beijing on a small scale, then in Liaoning province in northeastern China. The tobacco, which could resist infection by the cucumber and tobacco mosaic viruses, was released to farmers in 1992, making it one of the first transgenic crops to be commercialized in the world. It was planted widely on China's tobacco fields until the prospect of rejection in export markets forced the China National Tobacco Corporation to end production. Chen's laboratory at Peking University continued to develop transgenic plants while turning out a number of today's prominent or up-and-coming scientists and entrepreneurs. Though Chen still maintains his lab, he has long focused on shaping biotechnology policy, first as a Vice President at Peking University, then, since 2002, as President of China Agricultural University. He has founded many companies focused on plant and biomedical applications of biotechnology, and served for fifteen years on the National Biosafety Committee, the body responsible for evaluating safety and deciding approvals of transgenic crops.

Fig. 5.6 Top: Field trials of China's first transgenic crop, a virus-resistant tobacco variety. Bottom: Transgenic petunias developed at Peking University. Photos courtesy of Dr. Chen Zhangliang, China Agricultural University. (A color version of this figure appears between pages 72 and 73.)

Box 5.4 Protecting Cotton Harvests: Prof. Guo Sandui and the Chinese Academy of Agricultural Sciences

Born in a rural part of Zezhong County, Shanxi province, Guo Sandui spent most of his childhood among plots of corn and other grains, but very seldom encountered cotton, which later became the focus of his career. Guo spent the final years of the Cultural Revolution in classes at Peking University, which had partially resumed by that time, graduating just before the upheaval ended in 1975. He found his first job at the Institute of Microbiology at the Chinese Academy of Sciences, where he studied the *botchulism* bacteria, a frequent culprit in food poisoning. He also began investigating how the Bt gene (initially isolated from a bacterium) functions as a natural pesticide. In 1984, he transferred with his advisor, Dr. Fan Yunliu (see Figure 5.8), to the Chinese Academy of Agricultural Sciences, where his studies of the Bt protein continued. Just as the 863 Program was launched, Guo received an offer to work at the Pasteur Institute in Paris, supported jointly by the French and Chinese governments. His work in Paris on the structure and function of the Bt gene helped lay the foundation for his later contribution to the development of transgenic Bt cotton.

When he returned to Beijing, Guo's research group at the Chinese Academy of Agricultural Sciences joined several other laboratories already working to develop transgenic Bt cotton. By 1992, Guo's group had synthesized the Bt gene, with contributions from the Institute of Genetics and Developmental Biology at the CAS in Beijing. To introduce the gene into cotton, Guo initially used the *Agrobacterium* transformation method.

Once the Bt gene was synthesized, cotton was not the only target that researchers had in mind. The Bt gene was used at China Agricultural University for work on Bt corn and by Zhang Qifa's group at Huazhong Agricultural University (see Box 5.2) and others for work on Bt rice. At CAAS, researchers in Guo Sandui's group were also developing a Bt version of Chinese cabbage, but efforts were abandoned amid concerns that transgenic crops intended for human consumption might not be accepted by the newly emerging regulatory system.

Transgenic insect-resistant Bt cotton, on the other hand, gained widespread attention after conventional cotton crops had been increasingly devastated by pests, especially the bollworm, in the early-to-mid 1990s. Policymakers sensed an opportunity to realize the potential of one of the 863 Program's first fruits. Bt cotton plants were grown first under nets in field tests at the CAAS plantation in Langfang, Hebei province. Animal tests were also performed to ensure that the cotton oilseed did not produce any adverse reactions when eaten (Pray et al., 2006). Once Bt cotton had passed initial field testing, it was entered into large-scale field trials to assess its field performance in several different provinces (see Figure 5.7). While the details of commercialization will be left to later chapters, Bt cotton became the first transgenic crop in

Fig. 5.7 Guo Sandui examines Bt cotton grown on a testing field in China's Yellow River cotton growing region. Photo courtesy of Guo Sandui, Chinese Academy of Agricultural Sciences. (A color version of this figure appears between pages 72 and 73.)

Fig. 5.8 Dr. Fan Yunliu (left) and Dr. Jia Shirong (right), of the Chinese Academy of Agricultural Sciences. Photos by authors. (A color version of this figure appears between pages 72 and 73.)

China to reach farmers' fields on a large scale under the auspices of the biosafety system (although it was not the first transgenic crop to enter the market; transgenic tobacco was commercialized in the early 1990s).

Behind Guo's desk is a floor-to-ceiling panorama of a cotton field planted to Bt cotton developed at CAAS. Indeed, with early instruction from Dr. Fan, Guo was instrumental in the development and commercialization of Bt cotton. Along with his colleague at CAAS, Dr. Jia Shirong (see Figure 5.8), Guo played a central role in working with farmers to demonstrate its efficacy in rural areas of the Yellow River cotton growing region. Dr. Jia recalls how they had to win over many skeptical local stakeholders before the technology was widely adopted. CAAS has since expanded its research portfolio to include a wide variety of other crops.

During the 1980s, the scientists at CAAS also developed transgenic crops with other traits as well. In early 1993, Dr. Jia Shirong at the CAAS Institute of Vegetable and Ornamental Crops developed a variety of transgenic potato resistant to bacterial blight (Chen & Qu, 1997). Dr. Jia's laboratory is now part of the Biotechnology Research Institute, which replaced the Biotechnology Research Center in 1999. Dr. Fan Yunliu's laboratory worked on developing insect-resistant Bt rice, and other scientists at CAAS attempted to transfer the Bt gene into other crops, including Chinese cabbage.

As Bt cotton varieties were showing promising results in the field (see Chapter Seven), China's leaders redoubled their commitment to agricultural biotechnology research in the late 1990s. The funding allocated to research on rice, cotton, wheat, corn, soybean, potato, and oilseed has remained roughly proportional to planted area of each crop (Huang, Hu, Pray, & Rozelle, 2005a). Scientists began work on a broad range of transgenic crops, many of them unique to China. Among them were fungus-resistant cotton, insect-resistant soybeans, wheat resistant to Yellow Barley Dwarf Virus, potato resistant to bacterial blight and cotton beetles, and many others.

Perhaps the most extensive and controversial crop engineering project undertaken in China has been research to develop several types of transgenic rice. Research efforts have produced insect-resistant rice varieties transformed with CpTI or Bt genes (or both), as well as varieties resistant to bacterial blight, rice blast fungus, or rice dwarf virus, or tolerant to drought or salt conditions. The first varieties of insect resistant Bt rice in China were developed by a joint effort of Zhejiang University and the Zhejiang Academy of Agricultural Sciences. Later, collaboration between Dr. Zhu Zhen and his research group at the CAS in Beijing and rice breeders in Fujian province led to the development of insect-resistant rice using the CpTI gene.

In addition to insect-resistance, work on a wide variety of other traits in rice is well underway. Chinese scientists have developed blight resistant varieties transformed with the Xa21 gene, a gene that confers resistance in many species to infection by the bacterial pathogens belonging to the genus *Xanthomonas*. Isolated from a wild rice variety native to Mali in Africa, Xa21 was first described by a collaboration among scientists in China, France, Korea, and the United States (O'Toole, Toenniessen, Murashige, Harris, & Herdt, 2000; Song et al., 1995; Wang et al., 1999). A Chinese group led by Dr. Zhu Lihuang at the Institute of Genetics and Developmental Biology at CAS assisted with the cloning of Xa21, and together with Dr. Zhang Qifa's laboratory in Wuhan, worked to map the genetic loci for blight resistance as well as other traits (Chen & Qu, 1997). In 1998, scientists at IRRI introduced the Xa21 gene into indica rice varieties (Tu, Ona, Zhang, Mu, & Khush, 1998). Since then, both Drs. Zhang Qifa and Jia Shirong have succeeded in developing rice varieties with the Xa21 trait, and several strains have since reached field trials (Nao, 2005). In 2004, Dr. Zhang's lab reported the development of indica rice with both the Bt insect resistance and Xa21 disease resistance traits (Jiang et al., 2004). Dr. Zhang's current research also includes the development of drought-resistant and salt-tolerant rice by both transgenic and non-transgenic methods (Zhang & Luo, 1999; Yue et al., 2006; Hu et al., 2006).

Today, research is underway to apply transgenic techniques in a wide variety of other plant species, many of them of particular economic importance to China. Laboratory efforts to develop transgenic crop plants extend to cotton, wheat, corn, soybean, potato, oilseed rape, cabbage, and tomato (Huang et al., 2005a). China has also field tested varieties of herbicide-tolerant soybeans, insect- and disease-resistant corn, and disease-resistant wheat (Huang et al., 2005a). Transgenic poplar trees resistant to the gypsy moth are expected to reduce the costs of spraying pesticides in China's new green belts, which are designed to shield urban areas from dust and winds. Other laboratories in China are focusing on engineering crops to enhance nutritional or medicinal value, for instance by raising the content of essential amino acids or expressing vaccines in plant tissues. One laboratory at Peking University has expressed human insulin in tomato and potato, and another laboratory at the CAAS has developed potatoes that supply an edible form of the Hepatitis B vaccine (Lin, Hu, & Ni, 2004). However, at present, only a few crops are approved for commercial planting, including Bt cotton, virus-resistant sweet pepper, and virus-resistant and delayed-ripening tomato (see Chapters Seven and Eight for more information).

Although not the focus of this book, livestock applications of agricultural biotechnology are increasing as well. China's first transgenic animals have included pigs, goats, and mice that produce milk containing human proteins, and carp with increased body size (Fu, Hu, Wang, & Zhu, 2005). None of these organisms has been approved for commercial use. Microorganisms containing novel genes, such as transgenic yeast that can be used for food additives or bacteria that can remediate toxic chemicals, are also important areas of research in China. Animal cloning is also a major focus of research efforts at some agricultural universities (see Figure 5.3).

Non-transgenic Applications of Biotechnology

Although transgenic organisms account for a large and important fraction of China's agricultural biotechnology investment, their development has depended on a broad array of genetics, biochemistry, and molecular biology techniques. In the early years of the reforms, important work in non-transgenic applications of agricultural biotechnology took place at a growing number of institutes. For instance, several groups have used non-transgenic methods to develop virus-resistant varieties of tobacco, green pepper, tomato and other crops.

Functional genomics, or the identification of genes and relationships among them that influence their function, is growing increasingly prominent in China. A number of laboratories, including several key laboratories under the Ministry of Agriculture, are attempting to interrupt the sequences of all genes in a given crop with T-DNA (or mobile DNA sequences that can integrate into genomes) in order to identify their corresponding functions (Zhang, Guo, Chang, You, & Li, 2007; Chen, Jin, Wang, Zhang, & Zhou, 2003). Dr. Hong Mengmin's group cloned the rice waxy gene, which has since become an important focus of research efforts around the country (Chen & Qu, 1997). Such genome-level insights are important in molecular breeding applications such as marker assisted selection, which is widely applied in China's breeding institutes. Studies on the composition of the genomes of nitrogen-fixing micro-organisms are aimed at obtaining insights that could increase the efficiency of nitrogen fixation in plants.

Institutes across China have also been very active in improving on tissue culture methods to eliminate viruses from plant stocks or create large quantities of genetically uniform organisms. By the end of the 1980s, research efforts were underway at many institutes to develop techniques for improving major food crops. In China's Shandong province, the application of tissue culture to develop sweet potatoes resulted in yield increases in the field of up to 30 percent (Fuglie, Zhang, Salazar, & Walker, 1999). Chinese scientists were also among the first to apply anther culture in the breeding of haploid wheat and sugarcane varieties, and the regeneration of rice and soybean plants from protoplasts has also been successfully demonstrated in China (Hamer & Kung, 1989).

One recent addition to China's research portfolio has been the study of nutrient utilization in plants and fluctuations in nutrient levels in response to stresses, such as drought or high soil salinity. Studies by Dr. Zhang Qifa's lab at Huazhong Agricultural University (see Box 5.2) and others are currently working to identify the genes that regulate the flow of nutrients and enable them to respond to a particular stress. This fundamental understanding will aid in the design of crosses that maximize a plant's tolerance to water or nutrient stresses, with the prospect of lowering fertilizer and water requirements (Yan et al., 2006).

Major advances have also been made in understanding the molecular basis of agronomic traits that contribute to the high yield potential of rice and are critical for further increasing rice yield. Dr. Li Jiayang at the CAS Institute of Genetics and Developmental Biology has identified several important genes related to rice agronomic traits. Efforts by Dr. Li and others are complemented by studies on how regulation of individual gene function is coordinated at the whole-genome level. Dr. Li Zhikang at CAAS has taken a whole-genome approach and recently defined separate regulatory networks that can independently confer rice crop tolerance to drought or salt stress while maintaining high yield under normal conditions. This work will aid in the design of breeding strategies to enhance the tolerance of rice crops to water or salt stresses. These efforts also provide fundamental insights that aid in the development of stress-tolerant transgenic varieties.

Role of International Institutions

China's early biotechnology programs coincided with efforts by international donors to support further development of agricultural technology (and biotechnology) worldwide. One of the most influential in China was the Rockefeller Foundation's International Program on Rice Biotechnology. Though aimed at developing tools that would benefit all rice-growing nations, China's importance to rice production worldwide made it a natural candidate for support. Supplying nearly US $110 million over 15 years, the Rockefeller Foundation first provided support to developed world laboratories to apply emerging genomics techniques in rice (O'Toole et al., 2000). Scientists from developing countries were then supported to study in these labs, in anticipation that they would start programs in their own countries upon returning home.

The International Program on Rice Biotechnology had lasting effects on the development of rice biotechnology in China. In the early days of China's biotechnology program, the Rockefeller Foundation worked closely with the China National Center for Biotechnology Development to identify grant recipients. When only two of the original 20 grant recipients returned to China after studies abroad, the program shifted its focus to career scientists just starting out in China. Many of the recipients of these grants went on to become China's leading practitioners. Dr. Zhang Qifa, Dr. Chen Zhangliang, and many other leading scientists received funding to start their labs in China. In addition, two agricultural economists—Dr. Justin Yifu Lin and Dr. Huang Jikun—received Rockefeller support. Rockefeller's New York Office also assisted with equipment purchases in order to reduce costs and logistical difficulties of shipping to a still isolated China during the 1980s.

The World Bank also supplied significant funds to China, mostly for equipment upgrades in China's laboratories during the 1980s. In 1981, the World Bank provided a US$200 million loan to upgrade equipment and laboratory facilities, to which China contributed an additional US$95 million (Tang, 1984). Additional loans over the following years continued and expanded this support. These grants provided large number of laboratories across China with the basic equipment necessary to carry out molecular biology techniques, including laboratories at China Agricultural University, Huazhong Agricultural University, Peking University, and Zhejiang University.

Unique Features of Agricultural Biotechnology in China

Perhaps the most striking feature of China's agricultural biotechnology program is that research activities are predominantly funded by the government and driven by national policy priorities. The scope of China's publicly-funded research on transgenic organisms for agricultural uses is unparalled anywhere else in the world. While government budgets for agricultural research have declined in many countries, China has redoubled its investment and focused intently on biotechnology applications.

A strong tradition of state support for agricultural research has followed naturally from agriculture's economic importance. For most of history, China's economic output and tax revenues have relied on agricultural activities. Today, agriculture still accounts for a large slice of the economic pie, employing around 40 percent of the population and accounting for around 12 percent of gross domestic product (GDP) as of 2006 (*China Statistical Yearbook*, 2006; CIA, 2006). By contrast, the agricultural sector in the United States employs less than one percent of the total population and contributes less than one percent of GDP (CIA, 2006). China has a strong rationale for fostering the development agricultural technologies that reduce the burdens and help to increase the incomes of the rural population. This rationale is only strengthened by pressures to raise the competitiveness of the agricultural sector and improve prospects for food security, while sparing the country's environment.

The focus on applied research also distinguishes China's biotechnology programs from those in many developed countries. Although an emphasis on rapid returns to investment may have important advantages in a developing country, in the long run an overemphasis on applied research may undermine efforts by China's scientists to make fundamental breakthroughs that require sustained creative effort. Indeed, a United States National Science Foundation survey of applied research papers in Chinese journals published up to 1989 showed that 81 percent of articles sampled from biotechnology fields repeated research performed elsewhere (Hamer & Kung, 1989). Furthermore, as described in Chapter Three, forcing institutes to rely more on commercial sources of funding has diverted time and personnel from research agendas.

One must look only to recent history to explain the very low levels of homegrown private sector activity in China's agricultural biotechnology industry. In China's pre-reform centrally planned economy, the private sector was virtually non-existent—all enterprises were owned and run by the state. After the reforms began in 1978, the weak research capacity and insolvency of most state-owned enterprises left them poorly positioned to invest in research activities. In the case of agricultural biotechnology, ongoing reforms, especially in ownership requirements since the late 1990s, have created incentives for seed companies to consider research investments.

The emergence of China's agricultural biotechnology industry has not faced the public opposition that has emerged in other countries, particularly in Europe. From the beginning, the circle of scientists and policymakers who established China's biotechnology research programs encountered relatively little opposition or input from external influences. As most of China's rural population rarely encounters the research system firsthand, investments in agricultural biotechnology research have gone largely unnoticed among its potential users. Meanwhile, China lacks the type of consumer interest groups or an influential environmental lobby that have weighed in on the development of agricultural biotechnology—and particularly transgenic crops—in other parts of the world.

Taken together, these influences have shaped the size and scope of China's investment in agricultural biotechnology. Although China's research programs include both transgenic and non-transgenic applications, progress in developing transgenic crops, animals, and microorganisms has been especially rapid and extensive compared with most other countries. As of 2004, China's regulatory authorities had approved more than 1,200 cases or events (defined by the insertion of transgene in a specific background) for field release in trials (Huang et al., 2005a). Only a few of the transgenic crops in this category have been approved for commercial planting. Nevertheless, sustained increases in funding and support from the Chinese government have resulted in the emergence of an extensive capacity for agricultural biotechnology research. In the following chapters, we explore in more depth the potential and challenges involved in developing agricultural biotechnology beyond the laboratory.

6

From Lab to Field: A Changing Seed Delivery System

Successful laboratory development of an enhanced crop variety is only a first step on a long road to farmer's fields. The novel genes or genetic combinations must first be introduced into locally adapted varieties by cross breeding. Seed must then be produced in great quantities and delivered to farms. This process of scaling up for commercialization takes time and relies on a diverse collection of institutions within China's agricultural system, including research institutes, agricultural universities, seed companies, government extension offices, and farms. Whether and how rapidly new crop varieties reach farmers depends on how well these institutions perform and coordinate their functions.

This chapter explores how, against the backdrop of reforms in the seed industry and broader economy, a shift in the source of seed technology away from farms is affecting the development and adoption of agricultural biotechnology crops, and in particular, transgenic varieties. In developed countries, crop biotechnology is carried out primarily by large, integrated corporations that house research, production, and marketing departments within a single organization. By contrast, China's seed industry is relatively fragmented. Research, development, and sales functions are spread across a wide array of organizations. Many of them are struggling to survive and redefine themselves in the wake of reforms. Since the smooth operation of the seed delivery system is crucial for laboratory advances to reach farmers' fields, we examine more closely how recent developments in the science and China's reform agenda have affected its performance.

Seed Distribution in the Pre-Reform Years: 1949–1978

For thousands of years, China's farmers saved seeds at the end of each season for planting the following year, or obtained the seeds of superior varieties from neighboring farms. The general pattern of seed saving and sharing among farmers did not begin to change significantly until the middle of the twentieth century, when the government assumed more control over the seed delivery system.

As part of restructuring and collectivization in the agricultural sector during the 1950s, the new government established national research, breeding, seed distribution, and extension systems. Under the auspices of the Ministry of Agriculture, breeding research institutes, state-owned seed delivery agents, and a network of extension offices were set up at nearly every administrative level of government, including the national, provincial, prefecture, and county levels. At the provincial and local levels, research institutes were responsible for developing new varieties or introducing desired traits from imported varieties into varieties adapted to local growing conditions. During this time, all research and breeding activities were funded by the government, which in turn provided varieties free of charge to state-run seed providers (Fan, Qian, & Zhang, 2006). These seed providers arranged for seed to be multiplied on large growing areas designated as "seed production bases" before being transferred to farmers' brigades and teams. While this system supplied the necessary agricultural inputs, farmers had little control over planting decisions (Li, Li, Liu, Wang, & Jian, 1995).

Several aspects of the pre-reform seed delivery system are noteworthy. First, breeding research institutes depended solely on government funds to develop improved crop varieties. Since their success did not depend on farmer acceptance or any competitive advantage a new variety conferred, institutes with overlapping research agendas felt little pressure to consolidate their efforts, and research priorities did not necessarily reflect farmers' needs. Second, since research institutes' funding was virtually assured, the free replication of new seed varieties on farms did not adversely affect research institutes. Third, local seed distributors maintained a local monopoly, and as a result, they had few incentives to offer improvements in technology or services. Distributors also maintained close ties with local officials (either personal or monetary), which helped them maintain their monopoly status, ensure budget allocations, and support compliance with food security policies (such as grain self-sufficiency). Fourth, organizations that advised farmer planting decisions and provided regulatory oversight in the seed sector had no financial incentives to promote one variety over another, since all were disseminated virtually free of charge. Fifth, the path from lab to field involved many different organizations whose activities were at times only weakly coordinated. As a result, redundancy and inefficiency persisted.

Impact of Reforms

Since the early 1980s, China's reforms and opening to international markets have transformed the country's seed delivery system. Here we briefly describe how policy developments in China affected reorganization of funds and incentives within the seed delivery system, before exploring the implications of these changes.

In the 1980s and 1990s, the government implemented a series of reforms that encouraged all research institutes, particularly in the applied sciences, to grow more responsive to the market by reducing their reliance on government funds (Fan et al., 2006). In the agricultural sector, this transition was more gradual. The government was unwilling to turn over agricultural technology development completely to the private sector since its economic viability, particularly for technologies with social and environmental benefits, was considerably less certain. Still, many agricultural research institutes, particularly in applied fields, were forced to do their own fundraising. Many institutes began to redirect their efforts away from research and toward the marketing of well-established products, and a very limited fraction of commercial revenues were reinvested in research activities. The government system of guaranteed funding allocations for research institutes was modified to rely more on competitive grant applications, so that funds could be directed selectively towards national priority areas (Huang, Hu, & Rozelle, 2002b). However, full employment policies prevented many otherwise unproductive institutes from dismissing workers and closing down.

The reforms also included the restructuring of the nationwide government monopoly on seed distribution. For most of the pre-reform period, seed production and distribution activities remained unprofitable. Even as new (non-hybrid) varieties with higher anticipated yields were introduced, farmers' seed saving practices limited developers' ability to recover a profit. The central government also maintained some control over seed prices on grounds that high prices would inhibit adoption and threaten food security.

Only when hybrid varieties were introduced in the 1970s and 1980s did earning returns on seed variety development become a viable prospect. In the 1980s and early 1990s, state-owned seed distributors were gradually allowed to behave more like enterprises. Many were given increasing autonomy in management decisions and allowed to recover a share of their profits. Private sector competition began to enter the market for the first time. In 2000, a new Seed Law made it possible for private companies to sell seeds, provided that they could meet certain minimum asset requirements (Seed Law, 2000). Faced with an increasing competition, state-owned seed enterprises began to consolidate with other local and national-level companies. A new rule implemented in 2007 required that officials and government divest completely from local seed businesses, which was intended to further distance companies from local political agendas (Decree No. 40, 2006; Foreign Agricultural Service, 2006). The 2007 rule is also expected to mute official resistance to further consolidation, especially at the provincial and county levels.

Seed Development and Distribution Today

Although reformers intended for seed companies to grow more entrepreneurial and financially self-sufficient, actual changes have proceeded gradually and are far from complete. The government has occasionally backtracked on earlier liberalization policies amid concerns about food security and farmer access to seed. Some organizations have been able to adapt more quickly while others have responded more slowly, resulting in great regional diversity in the pace and extent of transformation. Today, it is difficult to describe a single channel through which improved varieties reach fields. Here, we describe the organizations and processes involved in seed delivery today as background for identifying remaining challenges to its efficient operation (See Figure 6.1).

Once researchers identify a suitable candidate for commercialization, the institute supports three years of multi-location field testing to assess agronomic performance. All new varieties are required to undergo agronomic trials, but the length and number of field trials also depends on any applicable biosafety requirements (discussed more in Chapter Eight). The results of agronomic trials (also called varietal registration trials) are then submitted to a commission of officials, local agricultural scientists, and seed experts who determine which varieties can be released at the provincial level. This process is known as varietal registration. Registration occurs at the provincial level if sales will be limited to one province; otherwise, a variety must be registered at the national level. Once a variety is registered, it is typically provided to seed companies, which are responsible for scaling up production.

Once an institute or company has received the green light to release a new variety, it is distributed through one or more of several channels. Sometimes an institute will market an improved variety by setting up its own seed sales company using funding from the government as well as outside investors (Fan et al., 2006). In other cases, a national-level research institute may provide its technology to provincial seed companies, either directly through contractual agreements or indirectly through provincial or local partner institutes. Local agricultural bureaus and independent seed sellers may also obtain improved seed from provincial or local institutes directly. Farmers can also buy seed directly from seed production bases. In principle, many paths exist to deliver new varieties to farmers, but in practice, diffusion depends on how well the incentives of developers and distributors are aligned.

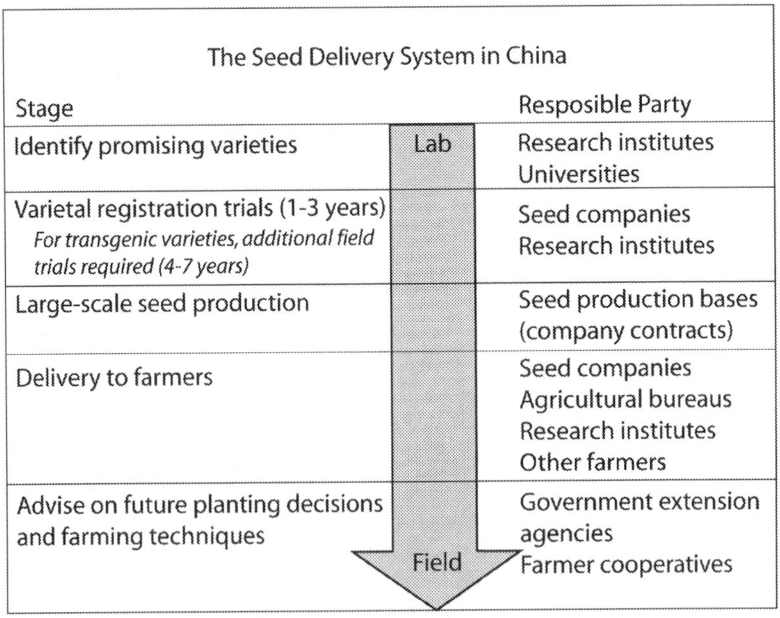

The Seed Delivery System in China		
Stage		Resposible Party
Identify promising varieties	Lab	Research institutes Universities
Varietal registration trials (1-3 years) *For transgenic varieties, additional field trials required (4-7 years)*		Seed companies Research institutes
Large-scale seed production		Seed production bases (company contracts)
Delivery to farmers		Seed companies Agricultural bureaus Research institutes Other farmers
Advise on future planting decisions and farming techniques	Field	Government extension agencies Farmer cooperatives

Fig. 6.1 An overview of the seed delivery system in China.

Seed companies may use several strategies to market their technologies. Together with local research institutes, extension agencies, and others, a seed company typically organizes field trials to demonstrate the potential of their seeds. Companies may also allow farmers to examine new varieties at trade fairs, or offer potential buyers a small amount of a new seed variety free of charge.

Challenges and Concerns

A shift in the source of seed technology from farms to laboratories has inspired new concerns and left room for potential gaps in oversight. Crops developed using agricultural biotechnology techniques must first be crossbred into local varieties, a process that requires time and trained personnel. In the case of transgenic crops, special measures to ensure human health and environmental safety add to the costs of commercialization. In many countries, most transgenic crops are developed by private companies with the intention of making a profit by providing desirable technologies to farmers. In order to protect their investments, these companies rely on strong intellectual property rights to prevent others from copying their technology. Finally, the costs of crop development are highly concentrated in the research stages with no guarantee that the technology will succeed. In developed countries, most of the investment is concentrated in large, integrated corporations that can afford high-risk investments in the face of uncertainty. However, the functions of China's seed development and distribution system are still highly fragmented, and most of the funding for research and extension activities is provided by the government. The following paragraphs explore whether or not these features of China's seed delivery system impose constraints on the adoption of improved crop varieties.

Collisions of State and Local Interests

Within the research system, the national and provincial academies of agricultural sciences have separate institutional mandates. At the national level, the Chinese Academy of Agricultural Sciences receives its budget from the national government, whereas the provincial academies rely mostly on local sources of public funds, such as taxes. Wealthier provinces are therefore able to support more research, and often set funding priorities independently of the center.

Where state-owned enterprises have historically dominated local seed markets, local officials and enterprises have tended to show a preference for locally produced varieties. The degree of state ownership and control varies widely by province and even by county. In some provinces, the traditional state seed supplier still maintains a monopoly. Traditionally, these suppliers have remained loyal distributors of locally developed varieties, and have had few incentives to shop around and deliver the best available technology to farmers. Further reinforcing this trend, the 1995 Governor Grain Bag Policy held provincial governors responsible for ensuring that a province maintained a stable grain supply at an affordable price (Crook, 1997). As a result, some provincial governments began to oppose policies they believed would threaten provincial self-sufficiency in seed production. This self-sufficiency policy may have had the net effect of ensuring supply, but inhibited adoption of advanced technology developed elsewhere in China, and was eventually dismantled. In part as a result of the decentralization of authority, China's seed supply is still highly diversified, which complicates national efforts to control quality.

A Weakly Coordinated System

In addition to the difficulty of managing a seed delivery system that spans multiple levels of government, the localized activities of agricultural research institutes, universities, and extension agents are also poorly coordinated. Separate and everchanging oversight makes coordination even more difficult. In the late 1990s, oversight of research institutes that had been subordinate to former industrial ministries was reassigned to the Ministry of Education, and many institutes were reclassified as full-fledged universities. Scientists previously engaged in the improvement and distribution of existing technologies had to redirect their energies to academic publications and teaching, disrupting an important function of the seed distribution system. In China, research activities were historically separate from education and extension, in contrast to the land-grant system in the United States, which integrates these three functions (Fan et al., 2006). In the absence of clear institutional roles and effective coordination among them, the mandates of China's agricultural research, education, and extension system often overlap in ways that prevent efficient allocation of personnel and resources (Huang et al., 2002b; Fan et al., 2006).

The Multiple Roles of Research Institutes

Reforms have also created competing mandates for China's research institutes. Faced with budget cuts, many institutes began to provide contract services to outside firms or reinvent themselves as profit-oriented businesses. However, pressure on bottom lines meant that institutes could no longer survive by turning over new technologies free-of-charge to seed distributors. In the absence of strong guarantees of compensation, many institutes found their energies redirected to activities with a dependable income stream. Sometimes these mission shifts had disastrous consequences—the China National Rice Research Institute in Hangzhou lost 10 million yuan after deciding to start manufacturing monosodium glutamate (MSG) in 1988, eventually bankrupting the factory (Fan et al., 2006). In other cases, pressure on staff to produce academic publications has conflicted with the institutes' efforts to develop better varieties. Over the last two decades, it has become common for researchers to assume many different roles at once, fragmenting missions and lowering productivity.

Fig. 6.2 Front door of a seed company located inside the campus of the Chinese Academy of Agricultural Sciences. Photo by authors. (A color version of this figure appears between pages 72 and 73.)

This trend is evident at many applied research institutes in the national-level CAAS. Many institutes were encouraged to undertake their own fundraising by selling their technology in the mid to late-1990s (see Figure 6.2) (Huang et al., 2002b). In 2000, 73 companies associated with CAAS had generated 120.5 million yuan (approximately 15 million dollars) from commercial activities, nearly half of their government funding allocation in the same year (Fan et al., 2006). Still, CAAS was expected to continue its policy of full employment, even under the new budget constraint. In the late 1990s, approximately 20 percent of the total CAAS budget (and 32 percent of its core funding) was used to support about 4,600 retirees (Huang et al., 2002b). Provincial institutes have faced a shortage of trained personnel as rising stars have left for more lucrative positions in industry. An increase in job opportunities following the reforms has created incentives for top researchers at local breeding research institutes to leave for national research institutes or other positions that offer higher salaries (Huang et al., 2002b).

Intellectual Property Protection

With the introduction of hybrid varieties in the 1970s, the notion that breeders should be compensated for their technology really began to take hold in China's agricultural system. Hybrid varieties require laboratory tools and specialized expertise to develop varieties with higher yields. The fact that hybrid seeds must be repurchased ever year in order to maintain higher yields has made it possible to introduce a viable compensation scheme. Developers of transgenic crops likewise require specialized laboratory tools and expertise, but unless a transgenic crop is a hybrid, farmers can use it just as they always have by reproducing it on the farm. In some cases, however, the properties of the original plant may be diluted over time. Still, developers are less likely to be able to obtain compensation if the state cannot guarantee protection for intellectual property.

Intellectual property protection is important both to facilitate technology transfer from research institutes to seed companies and to allow seed companies to recover licensing costs and any profits through sales. This protection also provides incentives for seed companies to invest in research and development of new hybrids. Previously it was common for research institutes and seed distributors to freely use crop plants developed by others as parents for producing improved seed. However, seed companies have been reluctant to pay for varieties they had previously received for free. When research institutes were not adequately compensated for their efforts, they sought alternative marketing channels or were reluctant to continue development. Moreover, as discussed later in this chapter, piracy of Bt cotton seeds eroded the revenues that companies recovered from sales of their product, and reduced incentives to disseminate enhanced varieties. Some question whether development of seed technology should rely on profit at all, since agricultural technology has traditionally been considered a public good.

The tradition of intellectual property protection in China is relatively young. The patent system was established in 1984, but implementation has largely remained weak, particularly at the provincial and local levels. In 1997, China's government adopted plant variety protection regulations, and in 1998, it signed on to an international convention recognizing plant breeders' rights (known by its French acronym UPOV). Although China has improved the strength and scope of regulations to ensure intellectual property protection over the past few decades, many challenges remain in the area of implementation (Huang et al., 2005a).

Dissemination of Transgenic Crops in China

So far, transgenic crops have been introduced into the Chinese market either by public research institutions or public-private collaborations involving both international and domestic companies. The transgenic tobacco varieties developed at Peking University were disseminated by the China National Tobacco Corporation until they were later withdrawn from the market (Pray, 1999). In the late 1980s and early 1990s, some international life sciences companies entered into collaborations with local institutes, but most of these early efforts met with little success. In addition to working with the Cotton Research Institute of CAAS in Henan province, the multinational company Monsanto also approached the Chinese Academy of Agricultural Sciences to discuss establishing a jointly sponsored rice biotechnology research institute. However, CAAS was unable to secure funds to supply its share of the capital, and concerns about intellectual property rights could not be resolved (Pray, 1999). The firm Ricetec, Inc. based in the United States worked with the Hunan Hybrid Rice Research Center to develop a market for hybrid rice in the Americas

(Pray, 1999). Later, other collaborative efforts emerged, but on a very limited scale. Restrictions on international participation in the seed market affected a collaboration to introduce Bt cotton in Hebei province, as will be described in the case of Bt cotton below. Since 2002, new regulations have prevented overseas investment in agricultural biotechnology research on the mainland. Foreign participation in joint venture partnerships has also been restricted to a minority share since 1997 (Keeley, 2003).

Another potential channel for disseminating products of biotechnology is through firms that belong to China's top breeding research institutes. For instance, Dr. Li Denghai, a leading corn breeder at the Laizhou prefecture's Academy of Agricultural Sciences in Shandong province, established a commercial hybrid seed breeding firm. Dr. Li's operation has successfully developed and marketed hybrid corn seed in Shandong province (Pray, 1999).

The Case of Bt Cotton

The case of Bt cotton illustrates how China's agricultural system might connect laboratories and farmers in order to accomplish distribution of the products of agricultural biotechnology research. To date, China has released many improved varieties, some of them introduced from overseas, and adoption rates have been high. The large number of research institutes that support development and dissemination of new varieties has helped to facilitate this process.

As scientists at the CAAS were developing the first strains of Bt cotton, China's cotton seed and procurement market was experiencing major structural changes. Until the mid-1990s, the government had issued procurement quotas for cotton and fixed sale prices, although once quotas were filled, the remainder could be sold at market prices. During the mid-1990s, quotas were often weakly enforced and then abolished by 1998, giving farmers more flexibility in their planting decisions (Huang, Hu, Fan, Pray, & Rozelle, 2002a). China's cotton seed market also changed with the entry of private seed businesses in 2000, which has affected seed prices, availability, and quality of customer service as competition has increased.

Only one transgenic crop, Bt cotton, has been marketed on a large scale. Initially, domestically developed varieties were slow to reach fields. It is interesting to compare the experiences of the two most prominent sources of Bt cotton technology in China, the Chinese Academy of Agricultural Sciences and the Monsanto Company. Though each case has distinct features, it is possible to discern from these two experiences the general path that a transgenic crop would follow from lab to field, and where potential problems could arise.

The case of Bt cotton is noteworthy not only because it represents the first introduction of a transgenic crop on a large scale in China, but also because the domestically developed (CAAS) variety competed directly with varieties introduced by a multinational company, Monsanto, in partnership with the Delta and Pine Land Company (DPL). Initially, there were major differences in the patterns of adoption of these two varieties. The Monsanto-DPL variety was adopted rapidly across a large area of Hebei province, while the domestically developed variety was accepted only slowly when it first reached the market. A number of factors may have contributed to the differences in the pace and extent of adoption in these early years.

Beginning in the mid-1990s, scientists and administrators at CAAS began to prepare for commercialization of their Bt cotton varieties. Field testing for Bt cotton under a variety of conditions began in eight provinces in 1994, and by the end of 1996, two CAAS Bt cotton varieties were approved for commercialization in nine provinces. However, adoption was initially quite slow—by 1997, 667 hectares were planted, and at least 10,000 hectares were planted in 1998 (Song, 1999). To promote the CAAS varieties, institute researchers traveled to localities to conduct additional field trials at the local level to demonstrate the technology's effectiveness.

Slow initial adoption of the CAAS varieties merits a close look at potential weaknesses in China's seed distribution system. At the same time that Bt cotton was developed and commercialized, reforms were encouraging research institutes to become more self-supporting. Local institutes that otherwise might have supported the effort to introduce the national CAAS varieties found their staff increasingly engaged in other activities, such as hybrid seed multiplication, trading of seed, and fertilizer or pesticide sales, to support themselves (Song, 1999). Finally, most farmers and local breeders doubted that the new cotton variety would actually work as promised. Despite the fact that new strategies for controlling the bollworm were desperately needed, government-run local seed monopolies remained reluctant to distribute a new, outsider technology (Pray, Ma, Huang, & Qiao, 2001b).

CAAS responded by setting up Biocentury Transgene Corporation, Ltd. (*Chuang Shi Ji*), a company based out of Shenzhen and partially owned by real estate developers, to manage the sales of Bt cotton seeds. Biocentury then licensed the technology to three provincial seed companies, and in 1999, CAAS Bt cotton varieties were grown on 100,000 to 120,000 hectares (Pray et al., 2001b). This approach was far more effective than earlier attempts to market Bt cotton through the CAAS institutional hierarchy.

In contrast to the CAAS experience, Monsanto encountered more difficulty in obtaining approval for its Bollgard brand of insect-resistant Bt cotton. In 1996, Monsanto, the Delta and Pine Land Company (DPL), and the Singapore Economic Development Authority formed a joint venture with the Hebei Provincial Seed Company. A new company, JiDai, was established, and it received national biosafety approval to market Bollgard in 1997 (Pray et al., 2001b). Ownership of the joint venture was initially split between Monsanto and DPL (67 percent) and the Hebei Provincial Seed Company (33 percent), but later adjusted to comply with 1997 ownership rules (Song, 1999). Prior to allowing the formation of the partnership, the Hebei provincial government organized field testing of three Monsanto-DPL varieties and two local non-Bt cotton varieties. In 1995, when the Monsanto-DPL variety 33B yielded 30 percent more than its local competitors, the provincial government decided to support JiDai's entry into the provincial market (Song, 1999). By the end of 1997, JiDai had completed the national biosafety approval process and began marketing the new variety.

The Monsanto-DPL varieties were approved initially for planting only in Hebei province until 1999, and then by 2000 for sale in Anhui and Shandong provinces, while CAAS varieties were approved for planting over a much larger area (Pray et al., 2001b). In 1997, the government reduced the maximum allowed percentage of foreign ownership in local seed companies to 49 percent amid concerns that domestic producers would lose control over their technology (Keeley, 2003). Despite these restrictions, adoption of Monsanto-DPL varieties was rapid. In 1998, the first year after approval, approximately 80,000 hectares were planted to 33B in Hebei province (Song, 1999).

Several aspects of the partnership seem to have favored Monsanto-DPL's success in Hebei province, especially in the early years. As part of the joint venture agreement, JiDai and its sales outlets would exclusively sell 33B, ensuring a large market share. Another factor that may have favored Monsanto-DPL's strong debut was the fact that the company had extensive funding and management expertise that was shared in part with JiDai's leadership (Song, 1999). Also, because Monsanto-DPL approached the provincial government directly for approval, it circumvented the cumbersome process CAAS faced in transferring its technology through multiple levels of government. Finally, in many respects, the initial Monsanto-DPL technology proved superior to other conventional varieties in early trials, winning trust at the provincial and local levels.

Since the introduction of Bt cotton technology, the situation has been less favorable for Monsanto-DPL. Not only did the partnership have to contend with widespread piracy of its seed, it further lost market share as adoption of CAAS varieties increased. CAAS varieties were generally cheaper, whereas the Monsanto-DPL Bt varieties were sold at a premium that included "technology fees" for seeds acquired directly from the company. Compared with CAAS varieties, the Monsanto-DPL Bollgard varieties were also slower to receive approval for commercial planting in additional provinces from Chinese regulatory authorities, as will be discussed in Chapter Eight.

Both Biocentury and Monsanto-DPL encountered difficulties in recovering revenues for the sales of their Bt varieties. Only a small fraction of the seed sold under the Bollgard brand was actually supplied by Monsanto-DPL; a number of outlets began selling fake varieties that were nearly indistinguishable in appearance from the pink-tinted Bollgard seed developed by Monsanto-DPL. Some observers have documented that fake seed was sold widely, even in provinces where Monsanto-DPL varieties were not approved (Keeley, 2003). Biocentury's efforts to obtain compensation for CAAS Bt varieties met with little success until the company was partially acquired by Origin Agritech Ltd., which has been able to collect royalties from some of the provincial seed companies. Still, CAAS itself has not benefited directly from the profits of Bt cotton sales. CAAS scientists maintain that rewards for successful technology development often take the form of increased grants and greater prestige for individual scientists and the institution. Perhaps the deeper issue is how to effectively distribute the roles of conducting basic research (which is far removed from new product creation) and applied research (which is more likely to be commercially self-supporting) among the organizations in China's research system.

Implications for Agricultural Biotechnology

The case of Bt cotton adoption suggests several barriers that may constrain the diffusion of new seed-based technologies to China's farmers. Rapidly changing industrial policies, weaknesses in intellectual property protection, and reforms that reduce overall support for agricultural research affected the experiences of both CAAS and Monsanto-DPL in their efforts to introduce new varieties into the market. These lessons, taken together with broader observations from the previous literature about the transformation underway in China's seed delivery system, offer some initial insights into the system's potential to effectively deliver the crops and other commodities developed with biotechnology techniques.

Although it is arguably advantageous that farmers are receiving a large slice of the benefits of China's vast public research program, low returns to developers may hinder research progress in the long term. Many scientists maintain that the rewards of commercialization are not large enough for research institutes to promote the rapid and effective distribution of newly developed crops. If China's policymakers want research institutes as well as companies to thrive, stronger measures will be needed to ensure that the technology's creators and distributors reap an adequate share of the benefits.

The level of intellectual property protection may further influence the willingness of companies to market improved seed technology. Strengthening enforcement of intellectual property rights would help to ensure that scientists and breeders are compensated for the new knowledge they create. Though efforts have had some success, as in Biocentury's efforts to prosecute those who copied Bt cotton seed, overall enforcement is still weak and uneven. Weaknesses in intellectual property protection adversely affected both Biocentury and Monsanto-DPL and their partners, demonstrating the debilitating effects on domestic and foreign seed providers alike. However, there are signs that protection is growing stronger. Also, as more and more Chinese scientists develop intellectual property that they want to protect, they are more likely to advocate for stronger protection.

As the case of Bt cotton demonstrates, the pace of adoption also depends on the alignment of local interests involved in promoting and delivering seeds to farmers. Some observers have suggested that resistance from agricultural input industries with a stake in the pesticide business further impeded Bt cotton adoption, since adoption was expected to reduce demand for pesticides (Huang, Hu, Rozelle, Qiao, & Pray, 2002c). Research institutes, government extension agents, and local seed sales outlets were among the collection of organizations that suffered from corresponding reductions in pesticide sales. It is therefore little wonder that Bt cotton was not well received by these members of local networks. Directly addressing these sources of resistance and discontent could remove to the diffusion of the technology.

Experiences with the introduction of enhanced varieties to date suggest that China's domestic seed delivery system has not been subjected to the same competitive pressures that have led to the emergence of highly integrated agricultural biotechnology firms elsewhere in the world. Indeed, as the case of Bt cotton shows, the still disjointed and rapidly changing seed delivery system in China does not offer a single or clear channel for moving improved varieties from lab to market. Such a channel is not only important for the market to operate efficiently, but also to ensure that farmers can identify and purchase high quality and high-yielding varieties. In its current state, China's seed industry is unlikely to be able to compete with the experienced marketing departments of overseas firms, which may be in part both a cause and effect of excluding foreign competition in industries such as transgenic crops. Furthermore, weak seed delivery channels may deter the types of foreign investment that could result in technology transfer opportunities for both Chinese and overseas seed developers.

Clarifying relationships among research institutes and seed companies spanning various levels of government could also help to hasten the diffusion of domestically developed crops. The Monsanto-DPL varieties were initially more successful than Biocentury's varieties in capturing their target markets, which suggests that moving a new crop through the layers of China's agricultural system can be cumbersome and inefficient. Biocentury was set up to circumvent the initially slow process of introducing seeds into markets and to manage the local distribution of the transgenic Bt cotton seed. Only since it was partially purchased by Origin Agritech Limited has Biocentury been able to obtain compensation for licensing its technology.

The potential effects of international competition, if and when it is allowed for transgenic varieties, could create considerable uncertainty for China's domestic seed businesses. The question of winners and losers grows even more important as China's seed distribution system is increasingly exposed to foreign competition and no longer enjoys a captive market. In the case of Bt cotton in China, farmers have been the main winners in the adoption of new technology, since public funds subsidized research efforts, and fixed government procurement quotas for cotton initially prevented prices from falling in response to increased supply (Huang, Rozelle, Pray, & Wang, 2002d). Benefits in terms of yield gains or reductions in pesticide use have accrued predominantly to farmers, as will be explored further in Chapter Seven. This distribution of benefits contrasts with most developed country markets, where the developers, seed companies, and farmers divvy up the profits, with farmers getting a significantly smaller percentage of the overall pie.

In striking the right balance of public and private support, China's agricultural policymakers would be wise to consider each crop and technology individually. Although many technologies may prove both profitable and socially beneficial, some crops may have significant broader societal benefits that are undervalued in the market. As an example, drought-tolerant rice may reduce large water requirements, but as long as water is virtually free, rice farmers may be unwilling to pay a premium for less thirsty crops. In cases where the government intends to promote an important shared societal benefit, policymakers will have to consider whether markets alone can be expected to deliver it.

Our discussion underscores that China's seed industry is not yet organized in ways that make it efficient and internationally competitive, although the situation is gradually improving. Despite the fact that China's seed industry has experienced consolidation since the late 1990s, no vertically integrated agricultural biotechnology firm spans the entire production pipeline in China. Given that one of the earliest reasons for launching China's agricultural biotechnology programs was to maintain near self-sufficiency in its own seed and grain markets, the fact that Monsanto-DPL was able to quickly capture a large share of the market must have been quite alarming.

Instead of moving ahead with the commercialization of transgenic varieties, China may be waiting to ensure that its own domestic seed delivery channels and support services can hold their own in China's large and rapidly evolving market. If more transgenic crops are approved, China's government will have difficulty preventing the entry of varieties developed overseas. Without an efficient seed delivery system and a more favorable commercial environment, China's domestic developers may be ill-equipped to compete. The pace of reforms thus far suggests that it may be several years at least before China's seed industry will be ready.

7

China's First Transgenic Crops: Farm Level Impact

Though a new crop variety may show promise in the laboratory and contained field trials, its success in farmers' fields is far from certain. Even the most sophisticated biotechnology techniques can prove fruitless if a variety is not well adapted to field conditions. Altering the genetic makeup of a crop by either transgenic or non-transgenic methods may further affect the complex web of relationships among organisms in the surrounding ecosystem. If a new variety does not offer clear advantages over its predecessors, farmers may be reluctant to adopt it, since enhanced seeds typically carry a price premium.

In this chapter, we focus primarily on the health, environmental, and economic impact of China's first transgenic crops at the farm level. Although proponents of transgenic crops claim that the benefits may be greatest for farmers in developing countries, available evidence from the field to date is very limited. China provides one of the first examples of large-scale adoption of a transgenic crop in a developing country setting. Since non-transgenic crops are thought to pose fewer risks and are generally less controversial, we do not discuss their impact in detail here.

Though transgenic crops were planted in 22 countries as of 2006, the experiences of other countries are not immediately transferable to the Chinese case (James, 2006). First, China's farms are generally much smaller, and its farming population relatively larger and poorer, than those in most markets where the crops have been adopted so far. Second, the distribution of the technology's economic impact may also be different since the country has a large population of farmers that benefit directly from cost or other savings, such as reductions in pesticide sprayings. Third, many of China's transgenic crops were developed through public research programs, which measure the technology's success in terms of social benefits as well as its profitability. Familiarity with China's experience allows for a more informed assessment of the potential and limitations of transgenic crops in developing economies that share some or all of these characteristics.

Only one transgenic crop, Bt cotton, has been planted on a large scale in China. Its introduction provides the only insights into farmer acceptance and field performance of a transgenic crop in a broad sampling of China's ecosystems and growing regions. Studies on the impact of Bt cotton at the farm level have been conducted in several provinces in China's major cotton growing regions by a team at the Center for Chinese Agricultural Policy (CCAP) at the Chinese Academy of Sciences. These studies were designed to quantify the economic and health effects of Bt cotton introduction. The same group has also conducted similar studies on Bt rice, which is now grown in farmer's fields on a limited scale as part of pre-production trials. Studies by other research groups in China offer additional insight into the interaction of Bt crops with surrounding ecosystems.

Challenges to Cotton Farmers

The introduction of transgenic cotton builds on a long tradition of cotton cultivation in China that began over 2,200 years ago. Historically, a mixture of Asian and African cotton varieties were grown in areas as geographically diverse as the western province of Xinjiang, the southern island of Hainan, and the North China Plain (Jia, 2004). In the 1870s, Upland cotton imported from the United States replaced previously grown varieties, since the latter's inferior fiber quality made it less suitable for processing. Cotton is China's most important cash crop and supplies the principle source of income for many smallholder farmers. The planted area today stretches roughly along the Yellow River and Yangtze River, with some fields planted in parts of Xinjiang, China's westernmost province (see Figure 7.1). From 2000 through 2006, China produced and consumed more cotton than any other country in the world (Foreign Agricultural Service, 2007).

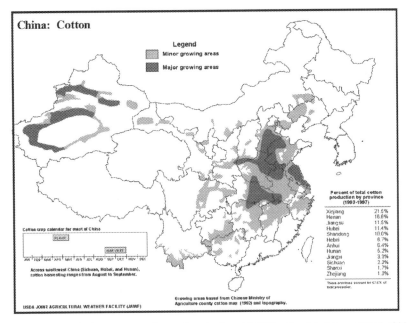

Fig. 7.1 Map of China's cotton growing regions in the mid-1990s. Reprinted from USDA (2006). *Weather and climate: Major world crop areas and climatic profiles.* (A color version of this figure appears between pages 72 and 73.)

For thousands of years, China's cotton farmers employed various strategies to curb pest outbreaks. In the 1960s, the first chemical pesticides to be imported from overseas greatly improved farmers' ability to control pests. Outbreaks of the cotton bollworm (*Helicoverpa armigera*) and other pests, including the tobacco budworm (*Heliothis virescens*), aphids, white flies, and the boll weevil, have long limited cotton yields in China (Jia, 2004). Integrated pest management techniques, in which pest control strategies are carefully tailored to a specific ecosystem, have also been applied widely (Huang, Hu, Pray, Qiao, & Rozelle, 2003).

In the 1980s and 1990s, cotton farmers in the Yellow River cotton growing region were spending increasing amounts of time and money to curb bollworm invasions. In some areas, the frequency of outbreaks is estimated to have doubled during the 1990s, which Huang et al. (2002c) suggest has followed in part from the intensification of crop production, a reduction in crop monitoring due to farm wage increases, and overuse of pesticides. Famers mainly responded by applying increasing amounts of highly toxic pesticides, such as chlorinated hydrocarbons (DDTs), until most were banned in the early 1980s. Farmers then switched to another class of pesticides known as organophosphates. When resistant pests emerged within a few years, organophosphates were replaced by less toxic pyrethoids in the early 1990s (Pray, Huang, Hu, & Rozelle, 2002). Resistant bollworms again emerged, and by the end of the 1990s, many farmers had already resorted to cocktails of organophosphates, pyrethoids, banned DDT from illegal sources, and other legal and illegal compounds to control the bollworm outbreaks (Pray et al., 2002). In the 1990s, farmers were applying more pesticides per hectare for cotton than for any other crop in China, with the exception of a few high value vegetable crops, which were only planted over small areas (Hossain et al., 2004). Meanwhile, bollworm outbreaks have continued to destroy a large and growing fraction of harvests, particularly during the late summer months. Had farmers not sprayed, Huang et al. (2002c) estimate that cotton yields would have fallen between 19.0 percent and 38.1 percent over the early to mid-1990s.

The health of farmers and ecosystems also suffered as a result of trends in pesticide applications. Farmers typically spend a significant portion of each season applying pesticides with small hand-pumped backpack sprayers and no protective clothing, and often return from the field covered in pesticides (Pray et al., 2002). Many farmers have reported feeling lightheaded or nauseous during or after spraying, and in some cases, have fallen seriously ill. From 1987 to 1996, an average of 54,000 poisonings and 490 deaths were recorded each year (Hossain et al., 2004). Soil and water quality have also suffered. Researchers have estimated that indirect costs due to the health and environmental damages of spraying pesticides, especially highly toxic forms, can exceed the private cost of purchasing them (Huang et al., 2002c).

Increases in pesticide purchases to counter the emergence of resistant pests have placed an increasing cost burden on farm households. In 1995, annual per hectare pesticide costs for cotton farmers reached US$101 (a total of US$500 million per year for the entire Chinese cotton market), a level much higher than for rice, wheat, or corn (Huang et al., 2003). The share of pesticide purchases in total agricultural input expenditures increased from 12 to 13 percent in the early 1980s to over

20 percent in the late 1990s (Huang et al., 2002c). The government extension system intensified its efforts to apply integrated pest management principles to control bollworm outbreaks in cotton fields, but with little success in reducing pesticide applications (Huang et al., 2003).

In parallel with the growing pest problem, China's cotton industry was undergoing reforms, some of which created additional challenges for farmers. State quotas for cotton production were removed in the late 1990s, allowing procurement prices to fluctuate according to market forces (Pray et al., 2002). Textile mills were allowed to buy directly from growers instead of only from state middlemen at fixed prices, resulting in downward pressure on procurement prices.

Increased pesticide costs and reduced yields, along with growing overseas competition, reduced farmers' incentives to grow cotton in the Yellow River cotton growing region in the early 1990s (Huang et al., 2002a). Bollworm infestations alone accounted for yield declines of 15 to 30 percent, and outbreaks in 1992 and 1993 caused economic damages of around US$630 million (Song, 1999). As a result, China grew more dependent on imported cotton, adding to the pressure on national policymakers to find ways to improve the outlook for domestic cotton production. One strategy involved expanding cotton cultivation in the water-scarce regions of the west by developing new fields and building irrigation systems. Another strategy consisted of increasing support for agricultural research aimed at developing technology that could change the outlook for cotton farmers, particularly in the Yellow River region where the damage was most acute.

Biotechnology for Smallholders: The Case of Bt Cotton

Severe and growing yield losses, rising pesticide applications, and the lack of a readily available solution were among the main factors that prompted the decision to commercialize several varieties of transgenic insect-resistant Bt cotton in the late 1990s (Huang et al., 2002c). Although the patterns of adoption are explored in more detail in Chapter Six, the extent of adoption is worth emphasizing here. By 2001, the area sown to both CAAS and Monsanto-DPL Bt cotton varieties had expanded to 1.5 million hectares, covering approximately 31 percent of the total cotton area (Pray et al., 2002). In Hebei and Shandong provinces, where the technology was commercialized earlier and the bollworm problem was particularly acute, adoption rates exceeded 75 percent by 2001 (Pray et al., 2002). By 2006, Bt cotton varieties covered 3.5 million hectares, or over 60 percent of China's total cotton area (James, 2006; Foreign Agricultural Service, 2007). Most of the adoption occurred on small farms of less than half a hectare in size (see Figure 7.2). The pace and extent of adoption suggested that farmers viewed Bt cotton as an attractive alternative to conventional varieties.

Farm level surveys and data analysis by Drs. Huang Jikun, Hu Ruifa, Carl Pray, Scott Rozelle and others at the Chinese Center for Agricultural Policy lend insight into why Bt cotton spread so rapidly. By interviewing 283 farmers in 1999 and between 350 to 400 farmers in 2000 and 2001, researchers quantified the impact of planting new cotton varieties on yield, household income, pesticide use, and farmer

Fig. 7.2 A typical village landscape in China's Hebei province. Photo by authors. (A color version of this figure appears between pages 72 and 73.)

health (Huang et al., 2002c; Huang et al., 2002d; Huang et al., 2003; Huang et al., 2005a; Pray et al., 2001b; Pray et al., 2002). Interviewers compared expenditures and health effects for randomly-sampled farm households planting transgenic Bt cotton with those planting non-Bt cotton. The 1999 survey was limited to a sample from Hebei and Shandong provinces. The 2000 sample included the original

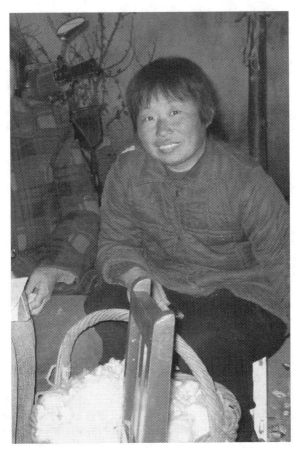

Fig. 7.3 A farmer in north central China's Henan province removes seed from Bt cotton. Photo by authors. (A color version of this figure appears between pages 72 and 73.)

surveyed group plus a sample from Henan province, and in 2001, farmers from Jiangsu and Anhui provinces were added. Hebei, Henan, and Shandong provinces are located in North China's Yellow River cotton growing region, while Anhui and Jiangsu provinces cover the eastern part of the Yangtze River cotton growing region.

In all three survey years, Huang et al. (2005a) found that farmers planting Bt cotton reaped substantial economic benefits. Reductions in both pesticide use and yield losses due to bollworm invasions contributed to a net increase in farmers' income (Huang et al., 2005a). In 1999, farmers saved an average of 20 to 33 percent of cotton production costs, although savings varied by location and variety (Pray et al., 2001b). Meanwhile, Bt cotton yielded an average of five to six percent more than conventional cotton on average (Pray et al., 2002). In general, gains persisted in 2000 and 2001, with the Bt cotton yield advantage averaging ten percent in 2001 (Huang et al., 2005a). Taken together, cost reductions and yield increases resulted in a net increase in total farmer household income of around US$500 per year (Pray et al., 2002). This sum could represent half a year's earnings for some farming families. By 2001, however, a fall in the price of cotton (due in part to the expansion of cotton harvests) passed some of the financial savings on to textile mills and consumers (Pray et al., 2002; Huang et al., 2005a).

In addition to contributing to increases in household income, reductions in the number and total volume of pesticide sprayings had substantial positive effects on human health and the surrounding ecosystem. In 1999, Bt cotton adopters needed to spray only 20 percent of the amount used by those cultivating non-Bt cotton to effectively control target pests (Huang et al., 2005a). Year-on-year total pesticide use on cotton was reduced by 20,000 tons in 1999 and by 78,000 tons in 2001, eliminating a quarter of all pesticide applications in China prior to Bt cotton introduction (Pray et al., 2002). However, reductions were not distributed uniformly, since pesticide use per hectare depends on the mix of pests in a given area (some of which may not be susceptible to the Bt protein) and can vary widely by province. For example, the red spider mite, a pest not targeted by Bt cotton, is the primary pest in Anhui and Jiangsu provinces, and thus reductions in spraying seem to have been more modest than other provinces where Bt cotton was adopted (Pray et al., 2002).

Research by Huang and his team also suggested that the significant reductions in pesticide use improved farmers' health. Surveys conducted in 2000 and 2001 revealed that only nine percent of farmers who exclusively planted Bt cotton reported pesticide poisonings, compared to 33 percent of those planting only non-Bt varieties (Hossian et al., 2004). Substantial reductions in pesticide applications have also decreased a persistent source of environmental pollution.

The studies by Huang and his colleagues have been widely cited as initial evidence that transgenic crops can deliver significant socio-economic and health benefits to small farmers in a developing country setting. The total financial benefits to farmers from adoption as measured by Huang et al. (2002d) are estimated at US$334 million in 1999 alone. Indeed, the financial benefits to farmers in the second year of adoption were enough to cover China's 1999 budget for agricultural biotechnology research (Huang et al., 2002d). Although the impact will inevitably vary by crop and by region, the CCAP studies on Bt cotton suggest that some agricultural biotechnology advances may offer sizeable economic, environmental, and human health benefits in China's rural farming regions (see Figure 7.3).

Bt Cotton and the Environment

Dynamic interactions between a cultivated field and the surrounding environment affect the long term viability of the agricultural system. When developing transgenic crops, it is important to consider whether or not adding a specific gene or combination of genes to a particular crop could affect this balance. If the introduction of a new transgenic variety (such as Bt cotton) causes a reduction in the population of one insect pest, its predators may starve while its prey thrives. When transmitted through the complex web of predator-prey relationships, these fluctuations could have unforeseen consequences. The realm of possible outcomes is limited by our current understanding of these complex relationships. Some of these effects are predictable, and are the subject of ongoing research.

We must, however, use caution when evaluating the environmental risks of adopting transgenic or other biotechnology crops to choose the appropriate case for comparison. In the case of Bt cotton, some risks may be due to the fact that the crop is transgenic, while other risks would be the same regardless of whether the pest resistance had been introduced by genetic engineering or conventional breeding techniques. Also, Bt cotton's economic, environmental, and health impact should be compared to the impact of the technology it replaces, which in China's case required large and increasing pesticide applications to preserve yields.

One of the major concerns surrounding the adoption of transgenic cotton is that its effectiveness in the field will not last. Indeed, this concern is not unique to transgenic crops, but it takes on new significance because the resistance is now an intrinsic property of the plant, rather than an external application. Some worry that as a result, transgenic material may be taken up by other organisms. Others are concerned that engineered DNA sequences (see Chapter Four) could spread in the wild, with potentially adverse consequences for genetic diversity. Some of this concern has focused on antibiotic marker genes, and has lessened in recent years with the development of alternative marker genes or marker-free transformation methods.

So far, according to studies by Dr. Wu Kongming and others at the Chinese Academy of Agricultural Sciences, there is no conclusive evidence that Bt cotton is developing resistance in the field. Still, scientists recognize potential concerns and have called for ongoing monitoring (Wu, Guo, Lv, Greenplate, & Deaton, 2003). The same group has taken measurements of bollworm populations in fields to aid in early detection of emerging resistant organisms (Wu & Guo, 2005). Laboratory experiments that involve feeding bollworms exclusively on Bt cotton have demonstrated that resistance does indeed develop after many generations (Jia & Peng, 2002). Bollworm resistance to Bt cotton is controlled by a pair of double recessive alleles. Therefore, the evolution of resistance can be limited as long as a population of insects containing at least one copy of the dominant (Bt susceptible) form remain in close proximity to breed with succeptible individuals (Liu, Tabashnik, Dennehy, Patin, & Bartlett, 1999).

In some countries, policymakers require farmers to plant non-Bt cotton on a small portion of their fields in order to provide bollworms and other target pests with a natural "refuge" that reduces their exposure to the Bt gene. Refuges are intended to increase the chances that any resistant organism will mate with those carrying the dominant (Bt susceptible) allele of the gene controlling resistance. On one hand, scientists believe that the fact that Bt cotton is planted on small plots in close proximity to other non-Bt crops in most areas of China reduces the need for refuges (Jia & Peng, 2002; Pray et al., 2006). On the other hand, in areas where Bt cotton fields are larger and adjacent to one another, such as in parts of western China, refuges may be needed.

The effects of Bt cotton on the broad spectrum of pests that exists alongside the bollworm in Bt cotton fields are important to forecasting long term ecosystem impacts. Studies by Drs. Wu Kongming and Peng Yufa revealed that levels of natural bollworm predators are significantly higher in Bt cotton fields compared with conventional fields (Jia & Peng, 2002). Insect communities are also comparatively more diverse, which suggests that using Bt cotton instead of chemical pesticides results in an increase in the diversity of insect life in fields. In addition, a decrease in the population of parasitic wasps that eat bollworm larvae has been observed. However, mirids (a particular family of leaf-eating insects not targeted by Bt cotton) have emerged as prominent pests in some years, which may have contributed to observed increases in the levels of spraying that followed several years after the adoption of Bt cotton (Wu, Lin, Miao, & Zhang, 2005). However, populations of cotton aphids, another prominent pest, are effectively controlled by the Bt cotton (Wu et al., 2005). These studies demonstrate that the introduction of Bt crops may have both positive and negative effects on farmer's ability to control pest populations. Nevertheless, verification that these effects are in fact the result of Bt cotton introduction and not due to other factors, such as fluctuations in weather or natural variability in pest populations, requires continuous monitoring over the long term.

As in other parts of the world, determining the nature and extent of these risks has been complicated by reports of emerging problems that are often not backed with scientific evidence. On June 3, 2002, the Nanjing Institute of Environmental Sciences and Greenpeace released a report entitled "Summary of Research on the Environmental Impact of Bt Cotton" that claimed the introduction of Bt cotton had harmed the environment (Xue, 2002). The paper did not present original scientific findings, but interpreted the findings of other scientists. Several of these scientists responded by alleging that their work had been taken out of context and misinterpreted (Wu, 2002; Jia & Jin, 2004). The Greenpeace report compared environmental impact on pest populations in Bt cotton fields with the impact in non-Bt fields that had not been sprayed with pesticides. Many scientists considered the pesticide-free non-Bt scenario an inappropriate basis for comparison given that it did not reflect the reality of China's cotton farming system. The report was also said to have misrepresented laboratory studies that showed resistance to the Bt gene could develop at high levels of exposure over many generations as evidence that resistant pests were already emerging in fields (Keeley, 2006). When the report was written, there was no documented evidence to suggest that the Bt gene's effectiveness against target pests had been reduced in fields.

Transgenic Rice: Prospects for Commercialization

With initial evidence of the farm-level impact of Bt cotton in hand, we now consider the potential impact of transgenic rice varieties. Rice is China's most widely planted staple crop. Although most prominently grown in the south, rice is planted in the north and consumed in nearly every corner of the country. From origins of domestication in central southern China thousands of years ago, rice has woven itself into local cultures and farming practices, many of which are unique to China or even to a particular group of villages. A total of nearly 30 million hectares are planted to rice in China, and hybrid varieties cover over half of this area. Indica rice varieties are primarily grown in south and central China over two growing seasons each year.

Japonica varieties are dominant in the north, where climate permits only a single growing season. Most rice crops are planted on small plots owned by individual farmers, many of whom earn a living primarily from rice-related activities.

China began developing the first transgenic rice varieties around the same time as Bt cotton in the early 1990s. Since then, the number of varieties and traits under study has grown steadily. However, only two varieties of Bt rice and one variety of blight-resistant (Xa21) rice have completed the final round of pre-production trials (an additional requirement since 2002) and are still awaiting final approval for commercial release. Insect-resistant Bt rice was viewed as an important research target mostly to reduce the large pesticide applications required over China's large rice-planting area. Although bollworms in rice fields could generally be controlled with a few pesticide sprayings, far fewer than in the cotton case, the total amount of pesticides sprayed on rice was greater than for any other crop in China (Huang et al., 2003). If approved and rapidly adopted, commercialization of Bt rice could expand China's area planted to transgenic crops from the fifth to the second largest of any country in the world.

Estimating Potential Impact

An initial study by Huang and his colleagues investigated the farm level impacts of two transgenic rice varieties in pre-production trials. One, GM Xianyou 63, contains the Bt gene to confer resistance stem borer and leaf roller. The other, GM II-Youming 86, is resistant to the stem borer using another gene that encodes Cowpea trypsin inhibitor (CpTI) (Huang et al., 2005b). Both varieties entered pre-production trials in 2001. Although only a small number of farmers actually grew the Bt rice as part of the field trials, a survey of a subset of randomly sampled field trial participants conducted by Huang et al. (2005b) provides first insights into the potential consequences of large-scale planting of transgenic rice.

The main benefits of adopting Bt rice were found to be reductions in pesticide use. Huang et al. (2005b) found that farmers only sprayed Bt rice fields 0.5 times per season, compared with 3.7 times for non-transgenic farmers. Corresponding pesticide expenditures were higher for the non-Bt varieties by eight to ten times (Huang et al., 2005b). Farmers adopting Bt rice spent only 31 yuan per season per hectare on 2.0 kg of pesticides, while non-adopters spent 243 yuan on 21.2 kg (Huang et al., 2005b). The mean yield of the insect-resistant rice was observed to be higher than non-transgenic varieties by 3.5 percent (Huang et al., 2005a). Huang et al. (2005b) also reported that instances of pesticide-related health problems were lower among the adopting farmers as well.

Challenges to Adoption of Transgenic Rice

Unlike Bt cotton, the challenges to adoption of insect-resistant and other transgenic rice varieties are more complex. Since rice is a major staple food crop, any unexpected problems would affect hundreds of millions of consumers. Although

Bt crops, including Bt corn and soybeans, have been accepted in the United States for human consumption, China's regulators have hesitated to move forward. Some European countries in particular have taken a more precautionary stance toward the adoption of the technology, as will be discussed further in Chapter Eight.

Insect-resistant rice in particular encountered a major setback when Greenpeace representatives sampled rice grown commercially in farmers' fields and found that it contained an unapproved transgenic variety (Greenpeace, 2005). This scandal was an embarrassment for China's inter-ministerial biosafety committee, which at the time was on the verge of approving insect-resistant rice for large-scale planting. It also drew negative international attention as it suggested weaknesses in the Chinese regulatory system and reignited debates about the prudence of moving forward with the technology. Discussion on adoption was shelved indefinitely and the offending varieties were traced to local seed companies, which were closed or reprimanded. Although many of the ensuing critiques reflected a renewed barrage of criticism toward transgenic technology in China, few observers carefully examined the reasons why farmers moved so quickly to adopt the technology. It appears that most were simply keen to realize its benefits, although they may have inadvertently slowed its adoption by doing so.

Other concerns related to transgenic insect resistant rice are similar to those raised by transgenic Bt cotton and will not be repeated here, but a few points are worth attention. In contrast to cotton, which is planted on small plots interspersed with other crops, rice is planted over large contiguous areas, offering few natural "refuges" for pests to avoid exposure to an insect-resistance gene (Zi, 2005). Another concern is the potential for rice, a cross-pollinating variety, to interbreed with its wild relatives and yield hardier, weedier varieties. Wild relatives are particularly abundant in China because it is home to one of rice's centers of origin.

Ongoing research on how to reduce emergence of resistant pests or other potential challenges will also help to preserve the effectiveness of new crops. Management can be expected to improve as biotechnologists gain more experience with the technology, and farmers and ecologists grow familiar with effective resistance management strategies. For example, scientists have already observed that using a "double gene" strategy (in which the expression of two independently-controlled gene segments confers protection against pests) can reduce the rate at which resistance develops (Jia & Peng, 2002). Another strategy involves designing an appropriate refuge or multi-cropping system to reduce the exposure of target pests. Evaluating the effectiveness of these approaches will also require a strong and comprehensive data collection and assessment capacity.

Potential Farm-Level Impact: Other Crops

In addition to Bt cotton, very few other crops have been approved for commercial planting in China. Around the same time that Bt cotton was commercialized, a tomato with extended shelf life, a virus-resistant sweet pepper, a color altered petunia, and a virus-resistant tomato were approved for commercial use in China by biosafety regulatory authorities. However, the transgenic pepper and tomato varieties did not offer significant economic benefits, and therefore companies hesitated to invest in their commercialization (Huang & Wang, 2002). Since 1998, China's biosafety authorities have only recommended approval of one transgenic crop, a virus-resistant papaya, although the number of crops in the approval pipeline has grown considerably over the same period. For instance, stress-tolerant rice, herbicide-tolerant rice, disease-resistant cotton, insect-resistant corn, quality improved corn, herbicide-tolerant soybeans, virus-resistant wheat, quality improved potato, insect-resistant poplar trees, and many other crops have all entered or completed field trials during this period (Huang et al., 2005a).

A number of agricultural biotechnology research laboratories are developing both transgenic and non-transgenic crop varieties that could reduce the impact of agriculture on the surrounding environment. Drought-tolerant varieties offer the potential to protect harvests in dry years as well as reduce overall water requirements, alleviating pressure on scarce water resources. Intensive cultivation has also left agricultural lands with high concentrations of salt, which can impede crop growth. Genetic changes in salt-tolerant varieties allow a crop's root network to adjust to take in water and nutrients even in the presence of high salt concentrations, which inhibit uptake of water and nutrients by plant roots. Many varieties, both transgenic and non-transgenic, have already entered field trials. Research programs are also developing varieties that require less fertilizer to maintain high yields. These varieties could help to reduce nitrogen-induced algal blooms and other adverse effects on China's agro-ecosystems (Yan et al., 2006). Many scientists are also applying transgenic and non-transgenic methods to improve nutrition or taste. Given that many people in China's rural areas suffer from vitamin deficiency, applications of biotechnology to raise the content of certain nutrients in dietary staples could provide an important health benefit.

Perhaps the most promising case for agricultural biotechnology is its potential to reduce the adverse health and environmental impact of intensive farming in China, while simultaneously enabling farmers to increase their earnings. The growing imbalance in rural and urban living conditions has put pressure on officials to identify new strategies to improve life in rural areas, with the hope of reducing tension and slowing the transition in the short term. The need for safer, more environmentally sound methods of crop production is substantial. Although transgenic crops pose at best a partial solution to some of these complex challenges, evidence from China thus far suggests that both rural households and the environment have benefited from adoption. Non-transgenic applications of agricultural biotechnology have shown similar promise, and may have even greater potential if transgenic applications remain controversial. The long term success of any agricultural biotechnology application, transgenic or not, depends not just on the technology itself, but on the ability of the research system to respond to the evolving challenges facing China's farm households and agro-ecosystems.

8

Biosafety and China's Regulatory Policy

Governments in most countries are entrusted with responsibility for ensuring that a country's agricultural sector supplies safe and high quality products to its citizens. To carry out this mandate, governments establish rules and enforcement agencies to assess the quality and effectiveness of novel crop varieties, food products, pharmaceuticals, chemicals, and many other items. In the case of crop varieties developed with biotechnology techniques, "biosafety" assessments carried out by one or more relevant agencies enable the government to establish and monitor the safety of newly developed crop varieties and products.

Health and environmental safety regulations serve many functions. They are primarily intended to prevent unsafe products from entering the market. However, they can also be an important enabler of a technology's diffusion by clarifying what criteria will be used to identify threats to public health and environmental safety, paving the way for products deemed safe to enter the market. In turn, developers use regulations to gauge how costly or difficult it will be to bring a product to the market. In performing these functions, regulations also establish the conditions under which products that originate abroad can enter the domestic market. If regulations are set in a transparent and consistent manner, developers, distributors, and consumers of a new product can act with greater confidence when making investment and purchasing decisions.

As the world's first transgenic crops edged closer to the market in the early 1990s, regulators were only beginning to consider how to assess their safety for environmental release and human consumption. These discussions emerged first in United States and Europe, but have since followed the diffusion of the technology to China and other early adopting countries. Many countries responded to concerns by drafting regulations that governed the management of biotechnology research and any resulting products.

Regulations can influence the development of a nation's agricultural biotechnology industry in several ways. Safety testing for a new crop can be expensive. By imposing costs at various stages in the development of a novel crop variety, regulations can alter the incentives facing the variety's developers, distributors, and consumers. The stringency of requirements and distribution of costs depend, in turn, on decisions within individual agencies charged with biosafety oversight. This chapter focuses on transgenic crops, since they have inspired new

V. J. Karplus and X. W. Deng, *Agricultural Biotechnology in China.*
© Springer 2008

biosafety concerns, and many governments around the world have implemented stricter regulatory requirements for them. Approaches to the regulation of transgenic crops have evolved differently across countries, driven by differences in history, public opinion, economic and trade incentives, and geographical considerations. This chapter describes how China's regulatory system emerged beginning in the early 1990s, and how it has been shaped by both domestic and international influences.

Global Context for Biotechnology Regulation

When the first transgenic crops were planted in farmers' fields, diverging views towards the technology had already begun to surface on opposite sides of the Atlantic Ocean. Most consumers in the United States were accepting or indifferent toward transgenic crops, while European consumers tended to be more skeptical. Concerned consumers, as well as industry and environmental interests, vied to shape decisions about how transgenic crops would be regulated in each nation or group of nations. In general, the European Union's policy has reflected a precautionary, wait-and-see approach, while in the absence of any demonstrated risks, the United States moved ahead with approvals. These approaches reveal differing risk perceptions as well as a lack of consensus over the way risk should be assessed. Though many agree that risk assessment should be based on "sound science," it is not always so clear what this means in practice, as stakeholders have not reached consensus on how to measure risk and what level of risk is tolerable.

When designing its regulatory system for transgenic crops, China, like many developing countries, was caught in the middle of a debate over which approach was most appropriate. In the early 1990s, when transgenic tobacco was first released in China, there were no formal regulations on the books, and most people were largely unaware of transgenic crops. The scientists directly involved with the first field trials elected to follow the rules used in early field releases in the United States. In the mid-1990s, as transgenic cotton was nearing commercialization, China adopted a more comprehensive, formalized framework for biosafety regulation. These regulations have grown more stringent since 2000, coincident with the intensification of international controversy.

This chapter describes how the Chinese government has interpreted the risks to human health and the environment in piecing together the country's regulatory framework for biotechnology. Beginning from the roots of the transatlantic controversy, it describes how risk management approaches have evolved in the United States and Europe and traveled with the scientists who returned to China to develop and introduce the country's first transgenic crops. The extent to which these practices have been formalized and implemented helps to explain why China, despite laboratory leadership, has experienced a slowdown in approvals of transgenic crops in recent years.

Divergent Attitudes on Transgenic Crops

As the first transgenic plants inched closer to market in Europe and the United States, a wide range of stakeholders vied to shape national approaches to evaluating the safety of crops produced with biotechnology. In the European Union, products made from transgenic crops were regulated separately from those derived from non-transgenic crops, on the grounds that the genetic engineering processes used to create them raised a novel set of concerns. In countries that adopted this approach, the developers of transgenic crops had to meet strict requirements before the crops could be disseminated by seed companies, grown in fields, or sold in stores. By contrast, in the United States, transgenic crops were regulated separately from conventional crops only if there were measurable differences in the properties of the final product. In practice, approvals in the United States proceeded more quickly since the responsibility for regulating transgenic crops was divided among existing regulatory institutions. In the European Union, by contrast, approvals of transgenic crop plants and seeds slowed to a stop between 1999 and 2002 while a regulatory framework was developed.

Several events at the end of the 1990s influenced the divergence of approaches to the regulation of transgenic crops. In the late 1990s, European confidence in government regulators was severely shaken by several food safety scares, particularly Mad Cow Disease in the United Kingdom. Most of the large multinational companies developing transgenic crops were based in the United States, and skeptical European consumers eyed their intentions with suspicion. Many Europeans questioned why they should accept the risks of transgenic crops in the absence of any direct consumer benefits. Indeed, the first generation of the technology was of greatest value to farmers and seed developers, and many of the latter were large agribusiness companies based in the United States.

Global outcry against transgenic crops intensified after the publication of several studies. In one case, researchers developed a soybean expressing a Brazil nut gene with high oil content, but later discovered that the transgenic soybean also expressed a common nut allergen that could provoke an allergic reaction in sensitive individuals (Nordlee, Taylor, Townsend, Thomas, & Bush, 1996). Greenhouse trials were immediately cancelled. However, many observers interpreted the incident as evidence of genetic engineering's potentially disastrous consequences. Others pointed out that the incident showed that regulations were effective in keeping unsafe products out of the food supply (Pinstrup-Anderson & Schiøler, 2000). Indeed, biotechnology techniques have also shown promise to enable scientists to detect and eliminate allergens in certain crops (Herman, 2003).

Once Mad Cow Disease had shaken public confidence in national food safety oversight, a series of reports by Dr. Arpad Pusztai further reduced the chances that transgenic crops would be quickly or widely accepted in Europe. At the end of 1998, a report by Dr. Pusztai and his colleagues claimed that rats fed on transgenic potatoes expressing a certain insecticidal protein from the snowdrop plant experienced significantly more intestinal damage than rats fed on non-transgenic potatoes supplemented with the same protein. Based on these observations, he concluded that the process of genetic modification itself introduced new, potentially harmful properties into the transgenic potato (Ewen & Pusztai, 1999). Instead of submitting a manuscript for peer review, which would have enabled other scientists to weigh in on the validity and merit of the results, he released his findings directly to the press. Dr. Pusztai's conclusions received widespread coverage, leading many uncertain consumers to reject the technology outright (Pinstrup-Andersen & Schiøler, 2000). Meanwhile, in the spring of 1999, the British Royal Society reviewed Pusztai's work and determined that no conclusions could be drawn from the data because the methods were flawed (Scott, 2003).

Outcry intensified when a group of American researchers announced that monarch caterpillars died at higher rates after eating leaves dusted with Bt corn pollen, compared with those fed on leaves dusted with non-Bt pollen or no pollen at all in a laboratory study (Losey, Rayor, & Carter, 1999). Although the methods were deemed scientifically sound, critics pointed out that the laboratory setup did not represent the actual field situation in several respects. Laboratory caterpillars were fed exclusively on Bt corn, while caterpillars in the field eat a diverse diet that at most only partially consists of leaves dusted with Bt corn pollen. Several follow up studies showed that in most parts of the United States, caterpillars were unlikely to be exposed to harmful levels of Bt corn pollen (Sears et al., 2001; Oberhauser et al., 2001; Hellmich et al., 2001). However, once the initial reports of harm to the monarch caterpillar had inflamed public opinion, subsequent studies on potential field impact mostly escaped the public eye.

The controversy and hysteria that followed the first reports on transgenic crops left little room for a reasoned dialogue on the risks and benefits of the technology. Criticisms on both sides of the debate were not only focused on the crops themselves, but on the fundamental distrust of the interests that advocated or opposed their adoption. Genuine concern over the actual risks of transgenic crops led many consumers to question why they should accept a risky, unproven technology, especially given that they would not receive direct benefits. Others argued that creating a transgenic plant was unethical because scientists were infringing on the realm of the natural and sacred by engineering gene transfers impossible to accomplish by sexual reproduction. Controversy also erupted over whether or not companies should be allowed to patent seeds, which were previously freely available and frequently traded among farmers (Pinstrup-Anderson & Schiøler, 2000).

The public outcry caught many developers off guard, most of them scientists who had labored for decades to create the new transgenic varieties. In laboratories, successes were met with genuine excitement and prompted executives to seek appropriate marketing and distribution channels. As part of efforts to enter the seed market, many agro-biotechnology firms sought to acquire seed companies that would enable them to establish new channels from laboratories to farms. Yet linkages that seemed useful to corporations were seen by others as alliances that assimilated longstanding institutions of the farming system, such as the family-owned seed company, into corporate conglomerates, while further marginalizing the voices of consumers (Charles, 2001).

Transatlantic Differences in Regulatory Approaches

The timing and extent of the controversy, as well as prevailing national approaches to regulation, influenced the development of regulatory systems for transgenic crops on both sides of the Atlantic. In the United States, responsibility for the regulation of transgenic crops was delegated to existing institutions. The United States Department of Agriculture (USDA) reviews field trial applications, gives seed developers authority to ship seeds from greenhouses to field trial sites, and reviews field trial results prior to commercial release. The Environmental Protection Agency (EPA) reviews applications for field trials larger than ten acres, and is further responsible for regulating all pest protected plants and assessing the impact on local insect populations and farmer pesticide use. The Food and Drug Administration (FDA) reviews studies that test for toxins and allergens and approves products prior to sale, while retaining the authority to remove any product from store shelves that later proves unsafe. Regulators in the United States do not require that foods made from transgenic crops be labeled, although many organic growers have opted to label their products as non-transgenic.

Many governments in the European Union (EU) have taken a more cautious approach to the regulation of transgenic crops, which is reflected in regulatory policy at the EU level. From 1999 to 2002, EU regulators restricted the importation and planting of transgenic crops until an appropriate regulatory policy had been implemented. This *de facto* moratorium prompted the United States to file a case with the World Trade Organization (WTO) claiming that EU regulations violated international trade rules by unduly restricting imports of transgenic crops in the absence of scientifically-demonstrated risks. Many government and citizens groups defended the EU moratorium on the basis of the "precautionary principle," which states that in the face of unknown risks, stringent regulation is justified. The case hinged in part on whether or not a precautionary approach clashed with WTO rules, which state that a country may restrict imports if they pose a scientifically demonstrated risk to the importing country's agricultural system (WTO, 1994). In 2006, the court upheld the United States' complaint, declaring that the EU moratorium constituted an unacceptable barrier to trade.

While the case was still under consideration, in 2002 the European Union enacted new regulations on the management of transgenic crops. These regulations built on laboratory-level oversight established in the early 1990s by Directive 90/219/EEC, which was comparable to the NIH guidelines developed in the United States in the early 1970s (Directive 90/219/EEC, 1990). The new regulations included Directive 2001/18/EC, which describes a rigorous set of criteria that must be met by transgenic crop developers prior to obtaining approval for environmental release (Directive 2001/18/EC, 2001). All member states must approve an application for environmental release before a crop enters the common market. European Community Regulation No. 1829/2003 outlines rules for placing food and feed products on the market, and along with Regulation No. 1830/2003 requires that certain measures be taken to ensure that transgenic plant material can be traced through all stages of the production process, so that potential cases of contamination can be easily investigated. The regulation requires labeling of all products that contain more than 0.9 percent transgenic material in a single ingredient (Regulation No. 1829, 2003).

In practice, the EU regulations are much stricter than their counterparts in the United States. Farmers growing products for the European market must adopt methods that ensure transgenic and non-transgenic varieties do not mingle. This process can be costly and difficult to enforce. The associated costs would be distributed along the production chain, and developers may either pass part of the cost to consumers in the form of higher prices, or assume the costs themselves, which may be prohibitive. Although transgenic crops can be planted legally in many European countries, only seven out of 25 countries grow them, and even then the transgenic crop area in each of these countries does not exceed 100,000 hectares (James, 2006). In comparison, the United States has planted a total area of 54.6 million hectares, and China grows 3.5 million hectares (James, 2006).

Global Adoption of Transgenic Crops

As the first transgenic crops moved from laboratories to fields, developing countries contemplating investments in the technology were caught in the middle of an increasingly vigorous debate. On one side were mostly scientists, companies, and international organizations that advocated strengthening biotechnology research and hastening the introduction of transgenic crops in the developing world. Transgenic techniques, they argued, would enable scientists to address persistent agricultural and health challenges, such as locally prevalent viruses or rust diseases or nutritional deficiencies, more quickly and inexpensively than non-transgenic crops or techniques ever could. On the other side, many environmental non-governmental organizations, governments, and citizens' activist groups encouraged developing countries to reject transgenic crops. They warned that the crops were not yet proven safe, multinational companies would control the inputs from abroad, and poor farmers would be unable to access, afford, or benefit from them. Many developing country governments remain concerned that any transgenic crops grown for export would be rejected in non-adopting European countries, reducing the value and marketability of their agricultural output.

Several cases have demonstrated how governments and populations in developing countries are often suspicious of the motives of the technology's advocates. Zambia was the first of several African nations to refuse food aid on grounds that it contained transgenic material. During 2002, the Zambian government rejected shipments of food aid in the midst of a famine primarily caused by prolonged drought (Bohannon, 2002). As of early 2006, the ban was still in place in several southern African countries. Several African countries have since begun to develop their own regulatory frameworks. Zambia adopted initial biosafety legislation in April of 2007, which calls for the establishment of a biosafety authority and a scientific advisory committee (Malakata, 2007). Indeed, in many parts of the developing world, countries have only just begun to consider how transgenic crops might be regulated within their own borders. Without a regulatory framework in some form, developing country governments have little means for demonstrating to their constituencies and the international community that they have evaluated the risks according to their own standards and made responsible decisions.

China's Evolving Biosafety Regulatory Framework.

When scientists returning from abroad began to develop China's first transgenic crops in laboratories in the mid-1980s, questions of how to regulate them were not yet on the government's agenda. The number of labs and scientists was very limited, and only a small group of researchers understood how the technology worked. This concentration of expertise contributed to a situation in which scientists were accorded significant authority and discretion in deciding if and how to monitor the safety of China's biotechnology activities.

In the absence of national regulations, China's first biotechnologists decided to develop their own safety procedures. When Dr. Chen Zhangliang's group carried out the first field trials on transgenic tobacco and tomato in northeastern China in 1991, they attempted to follow the same field trial procedures used previously in the laboratory where Dr. Chen studied in the United States. The main purpose of the trials was to evaluate the agronomic traits of crops, such as yield performance, as well as the efficacy of the newly introduced virus resistance traits. The tobacco was released for commercial planting shortly after the field trials, a practice that would not be permitted under China's current regulatory framework. The China National Tobacco Corporation later decided to withdraw transgenic tobacco when pressured by an importing company in the United States (Macilwain, 2003). While rejection of transgenic crops elsewhere was taking place at least nominally on safety grounds, in China the decision to withdraw transgenic tobacco was based primarily on commercial concerns about its reception in overseas markets.

Meanwhile, as awareness of biosafety issues was growing internationally, China's government began to consider how to regulate transgenic crops in a more comprehensive and coordinated fashion (see Figure 8.1). In 1993, the State Science and Technology Commission, precursor to the Ministry of Science and Technology (MOST), was charged with drafting the first guidelines for conducting recombinant DNA research (Huang et al., 2005a). These regulations served a purpose similar to the guidelines developed in the United States in the 1970s to regulate biotechnology in the laboratory.

Around the same time, responsibility for developing a more comprehensive set of regulations for agricultural biotechnology was vested in the Ministry of Agriculture (MOA). A preliminary biosafety committee was set up, and charged with comparing regulations from several countries, including the United States, European Union, and Australia, before drafting Chinese regulations. The final regulations, issued in 1996, closely resembled the United States' model with several key differences. The committee decided that transgenic crops had to be evaluated on a province-by-province basis, since crops may perform differently in China's diverse growing regions. Also, a separate set of field trials were required every time a novel

Major Developments in the Biosafety Regulation of Transgenic Crops in China

— 1993   **MOST** issues the "Measures for the Safety of Genetic Engineering" outlining procedures for approval and safe use of transgenic organisms

— 1996   **MOA** issues the "Implementation Measures for the Safety Control of Agricultural Organism Genetic Engineering"
— 1997   Ministry-level Biosafety Committee established within **MOA**
— 1998   "Administrative Measures on the Research and Application of Tobacco Genetic Engineering" issued by the **State Tobacco Monopoly**

— 2001   **State Council** calls for stricter biosafety measures in the "Regulation of the Safety Administration of Agricultural Transgenic Organisms"
— 2002   **MOA** issues three regulations implementing 2001 State Council regulations
MOA issues regulations on the health safety testing of transgenic foods
SDPC, State Economic and Trade Commission, and the **Ministry of Foreign Trade and Economic Cooperation** issue "Guidelines for Foreign Investment" limiting foreign participation in transgenic crop development

— 2004   Variety management regulations passed, requiring safety certification prior to registration of new varieties
MOA issues guidelinies simplifying biosafety approval procedures for transgenic cotton varieties
"Administrative Measures on the Inspection and Quarantine of Transgenic Products" passed, and administered by **AQSIQ**

— 2007

Fig. 8.1 An overview of major developments in the regulation of transgenic crops in China. Based in part on Huang et al. (2005a), Huang & Wang (2002), and authors' interviews.

gene was introduced into a new variety, even if that gene had been tested before in another variety. This provision addressed concerns that the gene may have novel properties when incorporated into a different position in the genome, or produce substantially different effects in a novel background. Aside from these differences, the regulations embraced the general principle that transgenic crops, once approved, should be treated as substantially equivalent to conventional varieties.

By the mid-1990s, as research on transgenic cotton progressed, policy makers paid greater attention to establishing biosafety procedures. A formal biosafety committee was set up under the auspices of the MOA to review and approve applications for field trials and commercial planting. The regulations established four stages of development for all transgenic crops and defined appropriate measures for evaluating safety concerns before proceeding to the next stage (Huang et al., 2005a). These stages included laboratory development, contained field trials (see Figure 8.2), environmental release trials, and pre-production trials. If no safety concerns arose, crops were recommended for commercialization and either approved or rejected by the committee.

Among the first crops to move through the biosafety system were several varieties containing the CAAS Bt gene, which were approved for commercial planting by 1997. At that time, government support for biosafety evaluation was very limited; nearly half of the early biosafety budget of US$120,000 was used to support biosafety testing for the CAAS Bt cotton varieties, even though research on many other crops was underway (Huang et al., 2005a). After transgenic tobacco was removed from the market in the mid-1990s, it was never again considered for biosafety approval because of concerns about acceptance in overseas markets. By 1998, several transgenic crops had been approved for commercial planting, including virus-resistant and delayed ripening tomato, virus-resistant sweet pepper, and both CAAS and the Monsanto-DPL Bt cotton varieties.

As the global controversy over transgenic crops intensified at the end of the 1990s, China's regulatory authorities were reconsidering the adequacy of the existing biosafety framework for transgenic crops. A lack of global consensus left China with contradictory models upon which to base domestic policies. Meanwhile, the scientific community was losing its monopoly on biosafety decision-making, as ministries and international organizations representing environmental, consumer,

Fig. 8.2 Contained field testing facilities in the Chinese Academy of Agricultural Sciences in Beijing. Photo by authors. (A color version of this figure appears between pages 72 and 73.)

or trade interests acquired more influence over the regulatory process. China was preparing to enter the World Trade Organization in 2001, prompting policymakers to consider the consequences of liberalizing trade on agriculture as well as other sectors. Many observers speculated that China's farmers would be unable to compete on global markets, especially against products from large mechanized and subsidized farms abroad. International environmental organizations, such as Greenpeace, worked through domestic channels to oppose the introduction of transgenic crops. A number of European and Asian governments made clear that shipments of transgenic crops and food products would be unwelcome within their borders. Shaped by a new set of concerns and stakeholders, China's regulation of commercial approvals of transgenic crops changed abruptly in favor of a more precautionary approach.

The administration of China's biosafety regulations was subsequently reorganized to reflect more stringent oversight requirements. In 2001, the State Council, China's highest governing body, amended the 1996 MOA regulations, enacting provisions relating to biosafety and trade in products derived from transgenic crops (Huang et al., 2005a). The fact that the State Council issued the regulations substantially enhanced their clout. The Ministry of Agriculture responded by issuing three regulations in the spring of 2002 that described how the new rules would be implemented. In order to allow for producers in China and abroad to adjust, interim measures were implemented until the regulations took full effect in 2004. The existing biosafety committee was upgraded from the ministerial to the national level, but the MOA retained responsibility for its administration (Pray et al., 2006). The 2001 State Council regulation and three MOA regulations were formally enforced with the passage of MOA Directive No. 349 in early 2004 (MOA Directive No. 349, 2004).

The new regulations were much broader in terms of the scope of health and environmental safety testing required. An additional pre-production field trial stage was added to the existing two stages of field testing required prior to application for commercial planting (see Figure 8.3). Pre-production trials involve at least one additional year of large-scale planting, and are intended to assess both yield characteristics and environmental impact. The new regulations delegated responsibility for evaluating the food safety of products derived from transgenic crops to the Ministry of Health (Huang et al., 2005a). The committee, which is comprised of food safety, health, nutrition, and toxicology experts, evaluates the results of tests for allergenic, toxic, or other health-related properties.

One of the most important developments that accompanied the new regulations was the creation of an inter-ministerial committee made up of members representing a variety of ministries to handle final commercial approvals, particularly in cases where controversy over a crop's safety involved a wide range of interests, such as the environment, public health, and international trade. Represented ministries included the Ministry of Agriculture, Ministry of Science and Technology, the State Development Planning Commission (now the National Development and Reform Commission, or NDRC), the Ministry of Health, the Ministry of Commerce, the Administration for Quality Supervision, Inspection, and Quarantine, and the State Environmental Protection Administration (Huang et al., 2005a; Pray et al., 2006).

The inter-ministerial committee does not meet regularly, but is only summoned to consider major decisions, such as the commercialization of transgenic rice. Since the new biosafety regulations were developed in 2001, only one transgenic crop variety, a transgenic virus-resistant papaya, has been recommended for commercialization (Dong, Song, & Liu, 2007). Meanwhile, approval of transgenic rice varieties has been postponed several times to allow for further safety and environmental impact studies (see Figure 8.4).

Trade and Border Control

China's government has also developed rules to regulate the safety of transgenic seeds, crops and derived products that originate outside of national borders. A country's right to control its own cross-border traffic in biotechnology products provides a crucial point of intervention for regulators. Regulatory goals can be quite diverse, and may extend beyond minimizing human health and environmental safety risks to include other aims, such as protecting domestic industry from international competition.

In 2001, China passed regulations that established safety criteria for evaluating shipments of transgenic crops or derived products. The regulations also called for all products containing transgenic material to be labeled as such, but did not supply clear implementation guidelines, causing confusion among traders both in China and abroad. Soybean exporters in the United States worried that their shipments would be rejected in China since they could not adjust immediately to the new regulatory requirements. Meanwhile, transgenic soybeans were not grown commercially in China, so Chinese producers did not have to make such an adjustment. Acrimonious debate yielded to a compromise in which producers in the United States were allowed to continue exports for a limited period to allow regulators to clarify implementation procedures and for exporters to respond accordingly. Today, China imports transgenic varieties of soybeans and corn under these regulations. Although widely consumed in the form of processed food products, no transgenic varieties of these crops were grown domestically in China as of early 2007.

Regulations pertaining to foreign direct investment in the seed industry clearly display a preference for national developers. On April 1, 2002, the National Development and Reform Commission and the Ministry of Foreign Trade and Economic Cooperation (later renamed the Ministry of Commerce) published the Catalog for the Guidance of Industries on Foreign Investment, which prevents investors abroad from engaging in transgenic crop development within China's borders (Huang & Wang, 2002). Many have questioned the legitimacy of the ban, which covers both joint partnerships for research and development as well as direct introduction of transgenic seed into China's market for sale to farmers. Imports of commodities containing transgenic material are also subjected to greater scrutiny compared with those developed domestically, since imported commodities (as well as agricultural inputs more generally) must be approved by the MOA prior to importation.

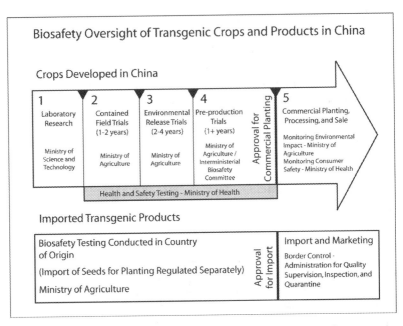

Fig. 8.3 A summary of biosafety procedures and administrative oversight for transgenic crop research, development, and commercialization in China. Based in part on Huang et al. (2005a) and Huang & Wang (2002). (A color version of this figure appears between pages 72 and 73.)

Fig. 8.4 Field trials of several strains of insect-resistant rice developed at the Chinese Academy of Sciences. Green patches are transgenic insect-resistant rice; brown patches are unmodified rice plants. Photo courtesy of Dr. Zhu Zhen, Chinese Academy of Sciences. (A color version of this figure appears between pages 72 and 73.)

The effectiveness of trade-related biosafety regulations depends on China's ability to enforce them at its own borders. National provisions for border control are implemented by the Administration for Quality Supervision, Inspection, and Quarantine (AQSIQ), an enforcement arm of the State Council. Increasingly, customs inspection checkpoints rely on inexpensive biotechnology techniques to screen shipments for transgenic plant material. China has not at present specified a minimum acceptable threshold for transgenic material—assessments call for a simple yes or no answer. Any shipment found to contain transgenic material that is not declared on the accompanying documentation is rejected. In practice, screening capacity varies widely, depending on the office's access to reliable screening techniques. Given that several high profile cases of transgenic material detected in unlabeled products have generated negative publicity, the potential for breakdowns in regulatory capacity at China's borders should also give cause for concern. The law does not include detailed information about enforcement or penalties, but states that violators should be prosecuted under other relevant laws (Decree No. 62, 2004).

What's Behind China's Regulatory Decisions?

The character of China's regulatory system and the institutions that administer it is a product of the early attitudes and approaches to the development of agricultural biotechnology in China. At the outset, a relatively small circle of scientists were the only ones who understood the technology, let alone considered how it might be regulated. Meanwhile, public opinion surveys have indicated that Chinese consumers know little about the technology. This asymmetry of information was reflected in the early regulatory decisions of representatives at MOST and MOA, which were met with little internal or public scrutiny.

The government's promotion of agricultural biotechnology was first called into question when the transatlantic controversy emerged in the late 1990s. The increasing concern surrounding the technology created an opportunity for the newly created State Environmental Protection Administration (SEPA) to vie for authority to manage biotechnology as part of its environmental protection mandate (Falkner, 2006; Keeley, 2006). In 2002, the SEPA collaborated with Greenpeace to publish a report alleging adverse effects of large-scale commercial planting of Bt cotton (described in detail in Chapter Seven). The report provided one channel through which SEPA and Greenpeace could influence biosafety oversight, which was mostly vested with the Ministry of Agriculture. International attention helped provide SEPA with extra leverage. However, in general SEPA's efforts to carve out a role in China's domestic biosafety regulatory process have met with little success. SEPA has instead been limited to developing regulations that implement the Cartegena Protocol, an international agreement that grants countries the right to reject imports of certain biotechnology products termed "living modified organisms" or "LMOs."

Biosafety and Commercial Concerns

Some observers have pointed out that, at least at the outset, biosafety regulation provided a means of protecting a vital sector of the economy from overseas competition and potential dependence on international sources of biotechnology. At the first meeting of China's biosafety committee, authorities addressed the question of whether or not to commercialize several varieties of Bt cotton. Some varieties had been developed by the Monsanto Company, while others had been developed by the CAAS (see Chapter Six for more information). Regulatory approvals for CAAS Bt cotton were granted more rapidly, and covered a much wider area, than initial approvals for Monsanto-DPL varieties. When the CAAS varieties, but not the Monsanto-DPL varieties, were later approved for planting in parts of the Yangtze River cotton region, regulatory authorities justified the decision on biosafety grounds, claiming that Monsanto-DPL varieties were less effective in those areas (Keeley, 2003). Their reasoning was that because they did not contain the CpTI gene (included in some CAAS varieties), Monsanto-DPL varieties would be less effective against late season pests, which are more of a problem in the longer growing seasons of southern parts of China. However, some observers have expressed little confidence in this stated rationale (Keeley, 2003). Another question concerned the suitability of growing the Monsanto-DPL Bt cotton varieties in the warm climate of the Yangzte region, since some of the candidate varieties had originally been adapted for planting in the United States (Keeley, 2003).

As experience with the technology and its regulation has grown, concern about losing market share to overseas entrants in agricultural biotechnology has abated somewhat. Monsanto-DPL and CAAS varieties are well established in the market (although as of 2004, piracy was still a major problem) (Huang et al., 2005a). However, to the extent that the biosafety regulatory system is still used to exclude unwanted varieties according to domestic policy agendas, it undermines confidence in China and abroad that the regulatory system is science-based and unbiased. In order to establish greater confidence in China's regulatory system abroad, regulators may be wise to address doubt about a variety's suitability or other concerns in a direct and transparent manner.

Consumer Attitudes in China

Compared to other countries described above, consumer attitudes toward transgenic crops have not played much of a role in shaping China's approach to biosafety regulation. Several surveys of consumer attitudes published in refereed journals have indicated that at least at present, most Chinese consumers are not averse to purchasing products derived from transgenic crops. Studies in Beijing and in other areas along China's booming eastern coast suggest that consumers show greater willingness to purchase transgenic crops than in other countries, although many interviewed showed little understanding of the technology (Huang, Qiu, Bai, &

Pray, 2006; Li, Curtis, McCluskey, & Wahl, 2002). However, a study conducted in Nanjing by Zhong et al. (2002) indicated that 20 percent of those interviewed had strong negative views toward transgenic crops, and almost all interviewees felt that foods containing transgenic materials should be labeled.

Since public opinion surveys are unable to accurately predict how consumers will decide in the supermarket, a study by researchers at the Chinese Center for Agricultural Policy and the United States Department of Agriculture looked specifically at how purchasing patterns changed once labeling requirements for transgenic crops were strictly enforced starting in 2003 (see Figure 8.5). The study suggested that labeling only mildly discouraged consumers from purchasing products containing transgenic materials (Lin, Dai, Zhong, Tuan, & Chen, 2007).

If Europe's experience is any indication, scandals or snafus, such as discoveries of biosafety violations or other unexpected events, could still undermine consumer trust. The fact that many consumers are not very knowledgeable about transgenic crops and have not yet formed a strong opinion suggests that future developments and media coverage could have an important impact on consumer preferences. Uncertainty about consumer acceptance may be part of the reason why China's leaders have hesitated to move forward with commercialization. If the dissemination of transgenic crops is not perceived as safe or equitable, domestic consumers may grow increasingly distrustful of the regulatory system and turn to international sources for guidance. Indeed, China's media itself is increasingly apt to look to international media sources for guidance on what topics merit coverage, with the result that the future of agricultural biotechnology in China will continue to be

Fig. 8.5 The label on the side of a jug of soybean oil in a Chinese supermarket. The label reads, "This product was derived from transgenic soybean but no longer contains any transgene product." Photo courtesy of Dr. Chen Zhangliang, China Agricultural University. (A color version of this figure appears between pages 72 and 73.)

influenced by developments abroad. Given that some top leaders have become more cautious in their approach to the regulation of transgenic crops, critical reports by the press may receive tacit approval for publication, with uncertain effects on the technology's future.

Maintaining China's Overseas Markets

Although agricultural products as a fraction of China's total exports have declined since 1980, maintaining export flows is an important goal of national trade policy. Regulations serve an important gate-keeping function by establishing the terms on which transgenic crops can enter and leave China's market. So far, China has not approved any transgenic food crops for export, making it difficult to predict the reaction in destination markets. However, concern that further approvals of transgenic crops in China could jeopardize exports to parts of Asia and Europe where transgenic crops are not accepted has left regulators reluctant to approve more transgenic varieties.

Evaluating Biosafety Regulation in China

How well does the China's biosafety regulatory system accomplish its mandate? From 1997 to 2003, China's regulatory system reviewed 1,044 applications to release transgenic organisms into the environment (mostly for contained field trials), of which 777 applications were approved (Pray et al., 2006). The applications spanned a wide variety of traits introduced into over 60 crops, animals, or microorganisms. For Bt cotton, which was approved before 1999, 30 new transgenic varieties were introduced by 2003 and 140 by 2004, suggesting that the process moves more swiftly for varieties transformed with genes that the biosafety system has already deemed safe (Pray et al., 2006). These varieties contain either Monsanto's Bt gene, the CAAS Bt gene, or the CAAS Bt plus cowpea trypsin inhibitor genes stacked together, which were all approved for commercial use in cotton in the late 1990s.

One measure of the regulatory system's success is to look at whether or not it has detected unapproved seeds, and if it has, how successful has it been at forcing illegal suppliers out of the market. In the case of Bt cotton, the Cotton Research Institute in Henan province registered a variety of Bt cotton without noting it was transgenic, and released it onto the market without first submitting it to the MOA's biosafety committee for approval. In field tests, this Bt variety was shown to be less effective at controlling late-season pests (Pray et al., 2006). The variety was subsequently recalled from the market, and the suppliers in Henan province were encouraged to distribute approved varieties. Although it had accounted for 20 percent of all Bt cotton planted in 2001, the unapproved variety has since been gradually replaced by approved varieties, in part because an alternative Bt gene construct (developed at CAAS) was readily available (Pray et al., 2006). Still, there is evidence that more crops, such as transgenic rice varieties, have slipped through the regulatory system unapproved (Zi, 2005).

The cost of meeting regulatory requirements can also be a major hurdle for would-be transgenic crop developers, but in China costs of regulation have been comparatively low. In China, the cost of bringing a transgenic crop to market is much lower than in India or in most advanced industrialized countries. The estimated total cost to a research institute for developing and marketing a new Bt cotton variety was about US$88,000, which includes US$75,000 for plant breeding and US$13,000 for compliance with biosafety regulations (Pray et al., 2006). In the cotton case, developers also saved most of the costs required for food safety testing, which are substantial for a crop such as rice. Private domestic seed companies that want to introduce the Bt gene into their own cotton varieties must also register their varieties. These companies are able to obtain the gene at a relatively low cost because already low royalty fees for CAAS varieties have not been well enforced (Pray et al., 2006).

In the mid-1990s, China's regulatory system allowed several transgenic crops to move from laboratories to farms, against a backdrop of strong government support for the technology. However, the future of transgenic crops in China looks far from certain. Even though regulations have grown stricter since 2000, releases of unapproved varieties suggest that China's biosafety system is porous in many respects. Moreover, negative public opinion toward agricultural biotechnology far from China has understandably left top leaders nervous about how to proceed with biotechnology approvals, both for economic and political reasons. Weaknesses in regulatory oversight suggest that the approval of any new transgenic food crops will require greater consistency and transparency, along with government and consumer confidence. Real or perceived weaknesses in biosafety oversight may prompt policymakers to hold off on the approval of transgenic food crops until they are confident that the system can prevent embarrassment or disaster.

9

Looking to the Future: Trends in Research and Rural Development Agendas

China's agricultural biotechnology investment has hardly occurred in isolation. Since the early 1980s, economic reform, opening to the world, and rapid economic growth have shaped China's broader national effort to develop expertise in the life sciences and biotechnology, as well as science and technology in general. These efforts further intersect with national development priorities, such as raising rural farm productivity and ensuring a safe, stable, and abundant food supply. In the last four chapters, we described some of the first results and remaining challenges associated with China's large national investment in agricultural biotechnology. We now turn to examine more recent trends in China's national policies, many of them extending beyond the agricultural sector. Throughout our discussion, we investigate how these trends might shape the future of agricultural biotechnology in China.

China's Science and Technology Renaissance

Since the early 1980s, China's support for science and technology has steadily increased. In 1995, at a conference on science and technology that rivaled major meetings in 1956, 1962 and 1978, the government redoubled earlier commitments with calls to revitalize the nation with science and technology (*kejiao xingguo*) (Suttmeier & Cao, 1999). Leaders called for total combined public and private spending on research and development as a percentage of GDP to reach 1.5 percent by 2000, and expanded public research funding accordingly (Suttmeier & Cao, 1999). It was at this meeting that the 973 Program was announced, which was intended to supplement the 863 Program, but with a greater focus on fundamental research.

Overall, growth in annual public funding for biotechnology and related applications has continued and even accelerated. From US$33 million in 1995, annual expenditures increased to US$104 million by 2000 (Huang et al., 2005a). By 2003, the budget had increased to US$200 million, of which US$120 million was allocated for agricultural biotechnology research (Huang et al., 2005a).

As government support for agricultural biotechnology research was growing rapidly, in 2002, President Hu Jintao's new government began to call for a rethinking of current economic growth strategies. Over two decades of growth at the national level masked inequalities along rural-urban, rich-poor, and east-west divides. Many state-owned enterprises found themselves burdened by insolvency and bureaucratic inertia. Reports of severe environmental degradation also suggested that economic growth was proceeding at the expense of human and ecosystem health. In response, the government announced its commitment to "scientific development," which was explained as a holistic approach to defining national development priorities that would reduce inequalities and improve livelihoods (*People's Daily Online*, 2006). In particular, greater emphasis would be placed on addressing development challenges in the rural areas.

A second goal of the new national policy was to boost the competitiveness of national industries by encouraging technological innovation. Despite sustained economic growth since the 1980s, China's leaders remain worried about long term dependence on foreign sources of technology (Segal, 2003). At China's National Science and Technology Conference in 2006, President Hu Jintao declared China would become an "innovation-oriented society" in the twenty-first century (Suttmeier, Cao, & Simon, 2006, p. 58). Announced at the same conference, the government's 15-year medium-to-long-term science and technology development plan (2006 to 2020) echoed this goal, with ambitious targets to raise the total research and development spending as a percentage of GDP from 1.3 percent (in 2003) to 2.5 percent, or US$113 billion (Suttmeier et al., 2006). The Chinese government aims for much of this increase to come from the private sector, in line with experiences in the United States and Europe, and has begun to develop new funding structures and performance measures that will aid in the realization of this goal (Wilsdon & Keeley, 2007).

The Evolving Rationale for Biotechnology Support

In the mid-1980s, biotechnology topped the national science and technology agenda, mostly for its potential to produce useful and economically valuable products. The 1990s saw this first goal joined by a second: strengthening support for original research to boost the nation's overall scientific stature and supply a source of fundamental breakthroughs that could lead to wholly new applications. As the downsides of economic growth grew increasingly apparent, rationale for biotechnology development was expanded to include its potential to alleviate environmental challenges and rural-urban inequality.

Since the beginning of the reforms in the early 1980s, support for fundamental research has grown, but has always lagged behind applied research. Funding for fundamental research as a percentage of total research funding continued to fall from 6.7 percent in 1993 to 5 percent by the late 1990s (Suttmeier & Cao, 1999). Modeled on the National Science Foundation in the United States, the National Natural Science Foundation of China (NNSFC) was created in 1986, primarily to support fundamental research. Its budget has expanded from US$21 million in 1986 to US$337 million in 2005, reflecting the growing importance of fundamental research on the national agenda (Chen et al., 2006; Wilsdon & Keeley, 2007). While

Fig. 9.1 The greenhouse of the Shanghai Institute of Plant Physiology and Ecology. Photo reprinted from Chen, H., Karplus, V. J., Ma, H., & Deng, X. W. (2006). Plant biology research comes of age in China. *The Plant Cell*, *18*(11), 2855–2864. Copyright 2003 by the American Society of Plant Biologists. (A color version of this figure appears between pages 72 and 73.)

average grants at the outset were not enough to start up a laboratory, they did enable existing laboratories to branch out into more fundamental work (Hamer & Kung, 1989). Projected increases in NNSFC funding of around 20 percent per year through 2010 will help to encourage more scientists to engage in basic research (Wilsdon & Keeley, 2007). The NNSFC was also one of the first, and remains among the few, grant-making organizations to use the peer review process to allocate research funds.

Efforts to recruit scientists to support the revitalization effort have targeted Chinese scientists trained overseas. Many have been encouraged to return with offers of generous funding packages and state-of-the-art laboratory facilities (see Figure 9.1). The success of recruitment drives is putting a small but growing dent in China's considerable "brain drain" challenge. Since 1978, over 580,000 scholars had gone overseas, but by 2002, only 150,000 had returned (Cao, 2004). According to the United States National Science Foundation statistics, of the 21,600 Chinese science and engineering doctoral candidates educated in the United States, 17,300 have remained (Cao, 2004). Today, an increasing but still very modest fraction of the total Chinese students educated abroad chooses to return to the mainland, many of them attracted by new or significantly restructured institutions.

Revitalizing Research: Reform and Experimentation

Over the last decade, China's science and technology reformers have sought to develop new structures and incentives that increase the effectiveness of research spending. One major effort was the Knowledge Innovation Program (KIP), which focused on reforming the CAS in ways that would make it more productive and internationally competitive in both basic and applied research (Suttmeier et al., 2006). Prior to the reforms, most institutes at the Chinese Academy of Sciences were overstaffed with non-research personnel and many aging, unproductive researchers. As in the case of the Chinese Academy of Agricultural Sciences (CAAS), over-staffing resulted in part from reliance on seniority rather than performance evaluation as a criterion for advancement. Quality of equipment was also variable across institutes, and most research programs were outdated. The main goal of the KIP was to create 30 internationally recognized research institutes by 2010, with five of them among the world's best in their fields (Suttmeier et al., 2006). So far, several leading institutes have emerged within the CAS system, including the Beijing Genomics Institute, the Institute of Neuroscience, and the Institute of Biophysics.

The KIP involved a major influx of funds and broad restructuring of the CAS. From 1998 to 2005, the number of CAS institutes was reduced from 120 to 89, either by converting institutes to commercial enterprises or reorganizing them to reduce duplication (Suttmeier et al., 2006). A larger share of funding was allocated to individual institutes directly, rather than to the CAS administration, giving institutes more flexibility to develop their own research agendas. A broad personnel recruitment effort was launched, drawing over 14,409 new researchers between 1998 and 2004, of which 68 percent were senior scientists under the age of 45 (Suttmeier et al., 2006). New appointees were to be evaluated early in their career to gauge productive potential. This evaluation approach places great pressure on new researchers, especially those engaged in fundamental research where new discoveries can take years, if not decades, to achieve (Ding, 2006).

In parallel with efforts to restructure the CAS, reformers began to experiment with new institutional models for encouraging original research. The National Institute of Biological Sciences, Beijing was one such model established to test the viability of an independent research institution administered outside the political and budgetary constraints of the Chinese Academy of Sciences and university system (see Figure 9.2). The government recruited promising scientists (all ethnic Chinese as of the end of 2006) with overseas training to set up their laboratories at the new institute. Each new principle investigator is granted a generous start-up package and guaranteed funding for the first five years. Continuation of funding is contingent on five-year performance reviews conducted by an independent and international Scientific Advisory Board. The institute's faculty is split between plant and animal sciences, with many faculty members that work on molecular-level applications in both areas. Although the institute only began operating in late 2003, its early publication record suggests that NIBS may become a workable model for supporting fundamental research in China.

Upgrading the Educational System

The rapid development of the life sciences in China is producing a new generation of scholars much larger than the previous ones. This new generation has grown up

Fig. 9.2 The front of the National Institute of Biological Sciences, Beijing. Photo reprinted from Chen, H., Karplus, V. J., Ma, H., & Deng, X. W. (2006). Plant biology research comes of age in China. *The Plant Cell*, *18*(11), 2855–2864. Copyright 2003 by the American Society of Plant Biologists. (A color version of this figure appears between pages 72 and 73.)

during the rapid reemergence of China's biotechnology industry. They are trained in the latest biological sciences techniques by a growing number of accomplished practitioners, and top students are often expected to go abroad for further training, either for full-time graduate study or as exchange students for several months or years. In China, most of the training happens at leading national and provincial universities, where increasingly capable academic departments offer students opportunities to engage in cutting-edge research. Graduate students also have the chance to work at the Chinese Academy of Sciences or new institutes such as NIBS, where they have access to equipment comparable to the best facilities in the world. Indeed, under the KIP, funding for graduate students at the CAS has been recently expanded, resulting in growth in the number of admitted students (Suttmeier et al., 2006).

Still, educating the general public in science and technology areas presents a considerable challenge. Only the top students have the opportunity to study at elite universities in China's major urban centers, and the affordability of tuition can be a major constraint on attendance. Among the general population, the level of familiarity with even elementary scientific knowledge remains low. China's top leaders have realized that if science and technology are going to play an important role in modernization, the public must possess some literacy in these areas. As a result, in the spring of 2006, the State Council announced a scientific literacy campaign. The campaign was intended to encourage more students to enter science and technology fields and raise the scientific literacy of the general public (Jing, 2006). Enhancing public education is viewed as important to realizing China's broader goals for raising national scientific competitiveness and sustaining economic growth over the long term.

Trends in China's Life Sciences and Biotechnology Research

By many measures, China's life sciences and biotechnology research are showing signs of progress. China's rank in terms of the number of publications included annually in the SCI Citation Index (a collection representing the most widely-read journals) rose to eighth in 2001, up from twenty-sixth place in 1985 (Cao, 2004). Injection of funds into facility upgrades and personnel training combined with reforms to increase overall efficiency have created an environment that at first glance resembles some of the most advanced research systems in the world. China's biotechnology programs—and the 863 program in particular—have led to several original and important applications of biotechnology.

International collaborations are further helping to strengthening research capacity on the mainland. Many principal investigators maintain laboratories in both the United States and China, and commute back and forth between them on a monthly or even weekly basis. These cross-border laboratories create natural paths for students to obtain work experience abroad and increase the reach of universities beyond traditional campus boundaries. Many of the collaborators abroad are of Chinese descent and have established their careers almost entirely outside of China. Those who have returned have pointed out that the rapid improvements in living and working conditions have made it an attractive career decision.

Despite major strides in developing China's life sciences and biotechnology research capacity since the start of reforms, most scientists and policymakers would agree there is still a long road ahead. Researchers working in the mainland have never been awarded a Nobel Prize, which has served as one rallying point for scientific revitalization. In several highly visible speeches, national leaders and researchers alike have declared that China is bound to win a Nobel Prize within the next twenty years (Cao, 2004). However, producing Nobel laureates is hardly a formula-driven enterprise. Some top overseas Chinese scholars continue to criticize the current system for not placing enough emphasis on creativity and basic research, while allowing elements of the old top-down, seniority-based system to persist.

Although imperfect measures of achievement, publication records also indicate that Chinese research system is still far outpaced by many developed countries. In 2000, scientists at Peking University (China's top-ranked higher education institution in the arts and sciences) published 1,105 papers in SCI-indexed journals, an impressive improvement but still only one-eighth the number published by Harvard University. In 2002, the average impact factor—a measure of the influence of a particular journal—of a Chinese journal article was 0.94, compared to 2.99 in Japan, 1.45 in Taiwan, and 1.24 in South Korea (Cao, 2004). Still, 0.94 was an all-time high for China, and current upward trends suggest that China could eventually catch up. There is also a danger that efforts to single out and establish top institutes will divert resource streams away from lesser known institutes at the provincial and local levels, a trend that may affect progress in plant breeding and other agricultural applications.

In the midst of rapid growth and the accompanying optimism about prospects for China's science and technology, history still plays an important role, one that will take persistence and patience to overcome. Erasing the ten year gap in education and training that resulted primarily from the upheaval of the Cultural Revolution has not been easy or straightforward. Still older history shapes the interactions of students and instructors. For a student to question the wisdom of his or her research advisor remains an unlikely occurrence, in line with old Confucian conceptions of teachers as final authorities. Laboratories often hire many students at wages considerably lower than most students abroad receive, constraining the amount of attention and resources a research leader can offer any individual student. Still, the situation is changing rapidly, and it is hard to know how long these constraints will persist. China has changed dramatically in the last thirty years, from a system where intellectuals were shunned to one that reveres them as the drivers of China's rapid development.

Rural Policy and Agricultural Biotechnology

In addition to the influences of China's broader national science and technology development efforts, China's agricultural biotechnology program is shaped by broader rural and agricultural development goals. As described previously, agricultural research remained historically separate from other types of scientific research, given its widespread economic importance and regional diversity of cropping systems and farmer needs. This gap has narrowed as more advanced laboratory techniques are applied to routine functions in crop selection and breeding. As a result, science and technology plans as well as national rural development priorities have been instrumental in shaping China's agricultural biotechnology programs. Here, we examine how shifts in rural policy have influenced the level, scope, and administration of research funding for agricultural biotechnology, with an eye to identifying important future trends.

China's broader rural development agenda suggests several aspects where agricultural biotechnology might address challenges facing the agricultural sector. First, as China's WTO commitments require that it open its agricultural sector to greater international competition, improved seed technology could raise yields or reduce input requirements, equipping China's farmers to more ably compete with counterparts abroad. Second, technologies that reduce costs, pest invasions, or adverse health effects could directly impact the welfare of China's rural citizens. Third, efforts over the last half-century to increase the productivity of agriculture have had detrimental consequences for the environment. Taken together, these specific concerns are also linked to broader government priorities, such as ensuring food security and political stability in rural areas.

In recent years, China's agricultural biotechnology development efforts have coincided with a resurgence of interest among national policymaking circles in addressing inequalities between rural and urban areas. Beginning in 2004, Chinese President Hu Jintao announced a series of initiatives aimed at improving conditions in rural areas. Many of the same goals were incorporated into China's Eleventh Five-Year Plan (2006 to 2010), which mentions policies to create "a new socialist countryside" (Lei, 2006). In practice, this effort includes targets for increasing agricultural spending, phasing out agricultural taxes, developing advanced technology, building new infrastructure, and eliminating tuition fees for rural students. Policymakers are hopeful that these programs will result in the integration of the rural countryside more fully into the national economy.

As a major part of the Eleventh Five-Year Plan, the sizable increase in government spending on agriculture (and agricultural research) is a particularly promising development. From 1980 to 1990, growth in agricultural research funding slowed substantially, coinciding with initiatives that encouraged institutes, particularly in applied research areas, to seek external funding. Although some of this spending supported biotechnology applications, substantial amounts were also allocated to other forms of agricultural research (such as agronomy, breeding, and soil science). Recently, many agricultural scientists and economists welcomed the reconsideration of research funding cutbacks, as they felt that the earlier approach had undermined broader, more fundamental agricultural research as well as research with primarily social rather than monetary benefits (Huang, Hu, & Rozelle, 2002b).

This increase in funding has also been accompanied by several attempts at reform and shifts in institutional responsibility (see Figure 9.3). Initially, funds allocated to agricultural biotechnology under the National High Technology Research and Development Program (863 Program) were administered by the China National Center for Biotechnology Development (CNCBD). However, in 2006 the responsibility was reassigned to the China Rural Technology Development Center (CRTDC), also under the Ministry of Science and Technology. The CRTDC, which was established in 1983, oversees the administration of a number of national science and technology programs focused on rural needs, as well as others to promote rural entrepreneurship, such as the "Spark" Program. Reassigning authority for agricultural biotechnology 863 Program grants has given the CRTDC greater control over more basic technology development efforts, with the intent that chosen projects would be more directly aligned with broader agricultural agendas. As mentioned in Chapter Five, more responsibility for administering agricultural biotechnology research grants has been delegated to the Ministry of Agriculture (MOA) in recent years, to ensure technology development efforts are matched to rural needs.

On the research side, policymakers considered whether or not the Chinese Academy of Agricultural Sciences (CAAS) should be subjected to institutional reforms similar to those undertaken at CAS. However, this proposal met considerable internal resistance from CAAS employees and from within the government, since the CAAS system provides an important source of employment that served a critical national need and had a long tradition of full state support. Instead of launching reforms of the type and magnitude pursued at CAS (which included revamping institutes, budgets, and personnel), a modest effort to improve research performance, for example by offering additional training to existing personnel, was pursued instead.

An additional institutional challenge has been the coordination of research agendas across a redundant and financially stretched research system. In terms of staff numbers, China has the world's largest agricultural research system (Huang et al., 2002b). For many years after reforms began, payrolls remained large (and included retired staff) while salaries stayed low, with the result that research projects were

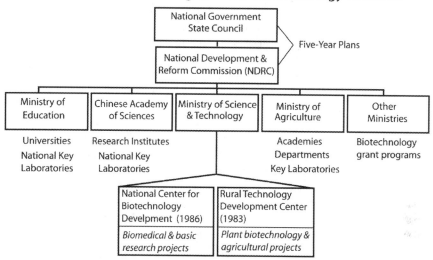

Governance of China's Agricultural Biotechnology Research

National Government
State Council

Five-Year Plans

National Development &
Reform Commission (NDRC)

Ministry of Education	Chinese Academy of Sciences	Ministry of Science & Technology	Ministry of Agriculture	Other Ministries

Universities
National Key
Laboratories

Research Institutes
National Key
Laboratories

Academies
Departments
Key Laboratories

Biotechnology
grant programs

National Center for
Biotechnology
Develpment (1986)

*Biomedical & basic
research projects*

Rural Technology
Development Center
(1983)

*Plant biotechnology &
agricultural projects*

Summary of Several Major Public Research Programs with Plant Biotechnology Components

Program	Year	Funding Agency	Description
National Key Technologies R&D Program	1982	State Dev. & Planning Commission (now NDRC)	Among the earliest program to fund biotechnology projects
National High Tech R&D Program (863 Program)	1986	Ministry of Science & Technology	Funds basic and applied research to meet national objectives
National Natural Science Foundation of China	1986	State Dev. & Planning Commission (now NDRC)	Supports basic research in the life sciences based on peer review
National Basic Research Program (973 Program)	1997	Ministry of Science & Technology	Supports basic research projects to meet national objectives
R&D Facility and Infrastructure Projects	ongoing	Ministry of Science & Technology	Part of Five-Year Plans, supports construction of Key Laboratories
Special Foundation for Transgenic Plants Research & Commercialization	1999	Ministry of Science & Technology	Supports transgenic plants research and commercialization

Fig. 9.3 Present governance of China's agricultural biotechnology research system and a description of several major public research programs. Based in part on Huang et al. (2005a), Huang & Wang (2002), and author's interviews.

often overstaffed or identical projects continued in parallel in the interest of full employment. By comparison, as of 2003, the private sector accounted for less than three percent of China's agricultural research budget, although its contribution to technology development was significant for its size (Huang et al., 2005a). These weaknesses still contribute to the difficulty in identifying promising researchers and coordinating agendas across many institutions.

The funding and institutional arrangements that prevail in China's agricultural research system represent improvements over 1978, but are still lagging compared to global standards. In terms of the percentage of agricultural GDP that is invested in agricultural research (a common measure of a country's commitment to agricultural research investment), China's investment was only 0.36 percent in 2000, and rose to 0.53 percent in 2005 (Hu, Shi, Cui, & Huang, 2007). For comparison's sake, developing countries averaged 0.53 percent in 2000 and developed countries reached 2.36 percent in the same year (Hu et al., 2007).

Even if increases were to be realized, the research system must be organized to take advantage of the increased resources. These institutional requirements might necessitate more aggressive reforms of the research system, establishing a robust peer-reviewed grant application process, increasing cross-institutional communication, and anchoring public technology investments in areas traditionally underserved by the market alone. Complementary policies to promote private investment would help to spur investment in profitable technologies, while helping policymakers identify which technologies require public support.

China's Science and Development Policy: Implications for Agricultural Biotechnology

This chapter has provided a broad overview of two important national goals that are shaping China's agricultural biotechnology research agendas: scientific leadership and rural development. Viewed against this backdrop, China's large (by some measures) public investment in agricultural biotechnology appears neither anomalous nor surprising. Here we examine several existing and plausible connections between the character of China's investment in agricultural biotechnology and its performance as measured against the stated goals of national policy.

First, the level of public funding for China's agricultural research programs, while modest by some measures, has been important in directing research toward targeted national priorities. Funds are more or less guaranteed over the window of a Five-Year Plan, and administrators consider stated policy goals when allocating funds. Since national priorities include a recently strengthened emphasis on "balanced" rural development, it is possible that future research efforts will focus on agricultural problems in underserved rural areas, or threats to food security. Bt cotton is widely cited as an example of a technology developed in response to an increasingly acute problem. However, merging the science and technology agenda with rural development priorities is likely to require greater communication and coordination among the relevant policy communities (most prominently MOA and MOST).

Second, simply because new technology is developed with public funding may not mean it will remain in the public domain or reach farmers at reduced cost. Some argue that the point of public technology development is to make its fruits available freely to all, since society is bearing the costs of development through taxation or other forms of public finance. However, the Chinese government currently grants an institution the ownership of any intellectual property generated using government funds. Policies also emphasize the cultivation of a competitive and innovative private sector, based in part on successful models in other parts of the world. However, with the emergence of private enterprise, China's agricultural biotechnology sector may begin to look more like its counterparts in advanced industrialized countries. If intellectual property rights are licensed to companies, new crop varieties that offer primarily environmental or other "public good" benefits may be bypassed in favor of the most profitable varieties.

Third, China has made strides toward reforming and modernizing its scientific enterprise (and supporting biotechnology in particular), but it is too early to gauge the effect that recent changes will have on China's long term competitiveness internationally. There are limited signs that China has begun to join the ranks of prominent contemporary contributors to basic research in plant biology. However, many of these achievements have come in labs that, although located in China, have strong linkages to overseas laboratories and collaborators. These linkages remain out of the reach of many laboratories (especially outside of major cities), but their directors must still compete with their internationally connected counterparts for national government grants. Indeed, science in China today is characterized by extreme pressure—both self-imposed and external—to publish in high profile journals. Cash prizes awarded for star papers further underscore the rather singular value China's system often places on publication record in evaluating scientific merit. Although the publish-or-perish mentality is certainly not unique to China, it is questionable whether, in the early stages, pressure to produce cutting edge results in short time frames is conducive to the development of a creative, flexible research system.

Beyond producing cutting edge innovators, the underlying aim of China's biotechnology programs is to cultivate a source of future economic growth and competitiveness. Ambitious growth targets for biotechnology product sales reveal these aspirations. However, as discussed in previous chapters, the future of biotechnology—agricultural and otherwise—depends not only on scientific achievement, but on the functioning of a broader set of institutions to deploy the technology safely and efficiently. Although agricultural biotechnology innovations, particularly transgenic crops, are only slowly reaching farmers' fields, this trend has thus far had little effect on broader support for agricultural biotechnology research in China. However, in the long term, the continued upgrading of research capacity is necessary, although it is unlikely to be sufficient, for China to reach the position of leadership in biotechnology to which its leaders and scientists aspire.

10

Conclusion

As the impetus behind modern industrial and agricultural revolutions, scientific advances have ranked among the most potent drivers of economic competitiveness and rising living standards over the last few centuries. Not surprisingly, governments under pressure to develop their nations have invoked this historical linkage to justify support for science and technology programs. Yet the challenges to acceptance of any emerging technology can be numerous, the rewards incremental, and the perils of implementing ill-conceived programs great enough that, should they come to pass, they could leave a country worse off than where it started. Thus the task of conceiving, establishing, and regulating a nascent science and technology enterprise capable of responding to national needs is a complex and delicate endeavor.

China's experience with agricultural biotechnology is a suitable case to explore these themes because its experience in building a modern, internationally competitive research program is only a few decades old. While cross-national comparisons or policy prescriptions are beyond the scope of this book, its pages have chronicled the emergence of agricultural biotechnology in China, with emphasis on its past and present challenges. Policymakers in other parts of the world, especially in countries with a mostly agrarian economy comprised of smallholder farmers, may find the experiences described in this book relevant to their own circumstances. Moreover, this book offers energetic young people working in agriculture, biotechnology, and related fields a glimpse of the broader picture in which their careers are unfolding.

The Role of History

Over the last few decades, China has built its biotechnology enterprise from the ground up, but China's long experience in agriculture and more recently in modern science and technology has underpinned its emergence and successes. Since China's first dynasties, advances in breeding, soil science, farming techniques, and understanding of seasonal variability have led to increases in the productivity of China's agriculture. Several of the most revolutionary advances included the iron plough, irrigation systems, and early-ripening rice, which enabled substantial gains in productivity on China's limited arable land. A commitment to experimentation,

V. J. Karplus and X. W. Deng, *Agricultural Biotechnology in China.*
© Springer 2008

observation, and broad adoption of superior methods was never entirely lost, even during years of chaos and upheaval between dynasties or during the Cultural Revolution. By the early 1980s, this commitment, combined with policy shifts and reforms, helped create an environment in which modern plant biotechnology could thrive.

Early twentieth century encounters with biological sciences research communities abroad also enabled the rapid expansion of plant biotechnology in China. In the 1920s and 1930s, a few internationally trained scholars who returned to China established the first plant biology research and training programs at universities on the mainland. Although this early experimentation dropped precipitously from the 1950s to 1970s, it had already laid the roots for modern plant biology and biotechnology research in China. Indeed, many early twentieth century scholars trained the generation that would help to revitalize and modernize the field in the 1980s. Without these early developments, efforts to strengthen China's research enterprise during the reform period would not have found such fertile ground.

Lessons from the Laboratory

The emergence of biotechnology laboratories as a prominent source of improved crops was hardly a foregone conclusion. When China's leaders redoubled support for science and technology in the early 1980s, the research system faced formidable obstacles. A limited number of trained personnel, aging and inefficient research institutions, and a lack of management experience hindered the development of a modern scientific enterprise. Developments since then cannot be attributed to any single policy or unified vision, but resulted from gradual experimentation once the constraints of the pre-reform period were loosened or discarded altogether. Decisions were originally taken in small circles comprised of the few who understood the technology, or held strong opinions about its importance for China's future. Yet the decisions of even the most forward-looking thinkers had to contend with institutional inertia, a chronic shortage of trained personnel, and limited funding.

The result has been a research system that is growing strong—especially in national flagship institutes that attract funds and the limelight—but lags in ways that could threaten future progress. Within the research system, several challenges remain. First, the rapid pace of development has generated new infrastructure and optimism, but placed a heavy burden on the individuals expected to deliver breakthroughs in short order. This type of environment compromises the thoroughness and originality of research. Second, the evaluation of both grant applications and research progress is still limited to a small group of scientists, a consequence of the program's relative youth, and makes truly independent peer reviewers hard to find. Third, funding also remains limited, even though China's research institutes may be able to economize by paying salaries commensurate with the lower cost of living in China and by sourcing equipment from local suppliers. These personnel and infrastructure costs will increase as China's economy and research environment grow more developed. Meanwhile, funding for life sciences research in China is still dwarfed by public and private sector investments in the United States and elsewhere. The private sector contribution, which has significant impact in many parts of the world, is still minimal in China.

Beyond agricultural biotechnology research, national investment has until recently placed comparatively less emphasis on strengthening traditional agricultural research fields. While support for cutting edge research in laboratories has the potential to push forward with advanced biotechnology research, it can also be detrimental because supporting technologies—particularly in breeding, soil science, and agronomy—are then proportionately underfunded. This trend is worrisome because these supporting technologies are essential to bridging the gap between the advanced technologies in the laboratory and practice in the field. For instance, in the case of transgenic crops, trained scientists are needed to introduce a gene construct into locally adapted crop varieties in order for the benefits of a novel gene to be realized in the field. If this step is not successful, the marginal benefits of any improvements made in the laboratory will be muted or lost. The fact that China's government is increasing its public investments in agricultural research and delegating more responsibility for agricultural biotechnology programs to ministries and centers with agricultural mandates is an encouraging sign. However, close coordination between biotechnology and agricultural agendas will require more than cosmetic rearrangements, but stronger communication among scientists and officials in both areas as well.

Many observers have asked if the uncertainties about future commercialization approvals of transgenic crops will affect China's broader investments in agricultural biotechnology research at the laboratory level. Although the previous chapters have focused mostly on transgenic crops, non-transgenic applications of agricultural biotechnology are well developed in China, and have demonstrated potential to increase the precision and efficiency of cross-breeding methods. Non-transgenic techniques do not tend to raise the same biosafety concerns associated with transgenic applications, but both rely on a similar body of knowledge and laboratory techniques. Since non-transgenic applications are less controversial and less likely to be subjected to stringent biosafety regulatory requirements, work in this area of research may enable important advances, even if the commercialization of transgenic crops remains on hold. This rationale may help to explain why China's broader agricultural biotechnology investment has continued despite the uncertain commercial prospects for many transgenic crops.

Lessons from the Field

Moving beyond the science, we ask whether or not China has developed the institutions necessary to deliver transgenic crops from the laboratory to the field efficiently and safely. Compared with most advanced industrialized countries, these processes are in many respects not yet well developed in China. This appraisal should come as no surprise, given that China is still undergoing reforms, and the institutions and infrastructure that would support and regulate the introduction of transgenic crops are not yet well established. The following paragraphs explore in detail the remaining barriers to adoption at the farm level. Throughout this discussion, the reader should bear in mind that progress is perhaps best measured in relative terms. Even in the markets of many developed countries, there is evidence that biosafety monitoring is not foolproof, and that the risks and benefits are unevenly distributed across the various stages of development. Broader ongoing reforms as well as targeted efforts are helping to lower many of the barriers to adoption in China, but at present these barriers may help to explain why more transgenic crops have not reached the market.

The benefits a new crop variety is expected to deliver to farmers—especially in response to a pressing problem—appear to be a major factor that affects its approval for commercialization and its adoption on farms. In the case of Bt cotton, increasingly severe bollworm outbreaks that reduced the benefits of planting cotton in China's eastern growing regions inspired the search and created demand for new technological solutions. The government could appear justified in its support for a technology that addressed an important need in rural areas. Future commercialization of other transgenic crops may similarly depend on immediate and demonstrated challenges on farms. Indeed, many of these challenges, such as land degradation and rising demand for more nutritious foods, are predicted to grow more acute during the twenty-first century. The development of a strong research enterprise capable of providing safe and effective technologies is a compelling rationale for public investment in agricultural research, including biotechnology.

Chinas' seed delivery system is still a weak link on the road from the laboratory to farmers' fields. Reliable, market-responsive seed delivery mechanisms are still lacking. In the case of Bt cotton, provincial and local seed companies hesitated to promote seed developed at CAAS, since it threatened the dominant position of their local varieties. However, this situation has begun to change, especially since 2000, as seed companies are integrating and consolidating at the county, provincial, and national levels. This ongoing transformation means that the partners available to research institutes for commercialization of transgenic crops are constantly changing, making it difficult to establish the lasting relationships that would ensure upstream developers are compensated for their efforts. Although plant breeders' rights are now established for all major crops, research institutes are limited in their ability to recover costs from the sale of improved seeds. This bottleneck provides a strong case for strengthening protection of intellectual property developed by research institutes and companies. Efforts to distance local government interests from seed company management may further help to foster investment in local seed markets. However, at present the underdeveloped commercial environment remains an obstacle to the distribution of transgenic and other value-added varieties.

A credible biosafety system will also be essential to sustain the development of transgenic crops in China over the long term. So far, Chinese consumers have not been significantly involved in, or in many cases even aware of, the government's investment in agricultural biotechnology. However, if additional crops, especially food crops, are commercialized, public confidence is likely to play an increasingly important role. Broad adoption will not occur if the technology is perceived as unsafe. Moreover, the government faces strong incentives not to push a technology when its safety is questionable. Even in the smallest villages, citizens are wary of unfulfilled promises and deeply skeptical of unproven technology.

International acceptance of China's biotechnology decisions will be critical, too. If negative publicity abroad over the discovery of unapproved transgenic varieties in fields or foods (such as in the case of Bt cotton or transgenic rice varieties) reaches Chinese consumers through increasingly open media channels and scientific communications, many are likely to be alarmed and alter their purchasing habits. Although new regulations have been developed at the national level, a track record of sound implementation at the local level will be critical to bolstering both domestic and global acceptance and ensuring the safety of the food supply.

China and the Future of Agricultural Biotechnology

In the laboratory, the successes of China's agricultural biotechnology program have been built on an age-old staple, agriculture, and a relatively new pursuit, modern science and technology. Without impetus for investment in both of these areas, it is unlikely that China's agricultural biotechnology industry, which lies at the crossroads of these two pursuits, would be as large or productive as it is today.

An eager reader might still ask if and when China will someday grow a broad array of transgenic crops on a commercial scale. Given the complex influences that bear on the answer to this question, the authors cannot pretend to offer clairvoyance. However, several observations reveal important concerns that will affect the prospects for commercialization. First, widespread adoption of transgenic crops will require corresponding advances and improved capacity for introducing beneficial traits into local varieties, either through transgenic or non-transgenic methods. Weak capacity for traditional agricultural sciences research could create a bottleneck that would slow the rate at which new varieties in laboratories reach fields. Second, the only transgenic crop widely planted so far was adopted in response to a widespread and recognized need. Transgenic Bt cotton was adopted because developments made by China's science and technology system (under the 836 Program) provided a means of avoiding the economic losses caused by increasingly severe invasions of the cotton bollworm. In the case of other crops such as rice, losses have not been felt on such a large scale. While there may be a potential market for other transgenic crops, in the absence of an immediate reason for approval, regulators may instead choose a wait-and-see approach. Third, the institutional linkages among the seed industry, research institutes, and government regulators at the local, provincial, and national levels are still undergoing reforms, and channels for disseminating transgenic crops are still somewhat redundant and not clearly defined or easily regulated. Finally, the biosafety system is not yet well established or transparent. Although the central government has pledged to strengthen the implementation of biosafety regulations, it may encounter difficulties in gaining the cooperation—or simply keeping track—of the provincial and local developers and seed companies. Until national authorities have confidence in the local monitoring capacity, the threat of international criticism of biosafety infractions may be enough to dissuade further approvals of transgenic crops, especially in the absence of a pressing threat to yields.

Much speculation has centered on whether or not China's support for agricultural biotechnology—and transgenic crops in particular—will allow it to soon surpass other countries in its development of a prominent, controversial technology. Some are concerned that a decision to commercialize transgenic crops in China, particularly a major staple crop such as rice, would constitute an irreversible global step, one that might be taken without international consultation. So far, these concerns have proven unfounded. Instead, the path of China's agricultural biotechnology development has been in large part shaped by developments abroad. Overseas commercial interests discouraged China's early efforts to market transgenic virus-resistant tobacco. China's regulatory authorities also relied on other nations for guidance in developing the country's first biosafety regulations. As global attitudes turned more precautionary, China's regulations evolved in the same direction. Overseas universities have trained many of China's top biotechnology scholars, and private foundations have provided guidance and support. Given the challenges China's nascent biotechnology enterprise faces today, the country's policymakers are perhaps more likely to seek rather than shun international support and consultation.

China also has a vested interest in managing the technology responsibly and in a manner consistent with international norms and practices. China's agricultural sector is still an important contributor to GDP, and employs a large (although declining) percentage of the population. Technologies that increase agricultural yields, reduce the need for fertilizers and pesticides, improve nutritional value, or offer other desirable properties could directly benefit the rural population. However, mismanagement could undermine health and environmental benefits, and tarnish the reputation of regulatory authorities as well as the technology's developers and distributors. As a result, policymakers face strong incentives to exercise caution when considering further commercialization of transgenic crops.

In step with research investments, China is emerging as an important contributor to plant biology and biotechnology fields internationally, as described in Chapter Nine. Over the last few decades, China has also taken an increasingly prominent role in transferring agricultural technology to developing countries, most prominently in other parts of Asia and more recently Africa as well. Many of these countries may find it more palatable to receive technology and advice from China, rather than rely solely on advanced industrialized countries for aid or technology, which may be ill-suited to smallholder farmers' needs or local growing conditions. In some countries, technology transfer from China has been accompanied by investment in research infrastructure and personnel training. If these efforts bolster the emergence of strong local research competency that enables recipient countries to address local agronomic challenges, it could rank among the greatest successes of China's agricultural biotechnology investment.

China's agricultural biotechnology programs are inextricably linked to national aspirations to advance science and technology and improve living standards in the rural sector. In order for science and technology to bear on individual lives and economic competitiveness over the long term, a national program must be able to accomplish two functions. First, it must be able to develop and deploy technologies that meet specific and tangible needs. Second, it must offer the fundamental knowledge base and flexibility required to develop appropriate technologies as new needs arise. These dual objectives may explain why China has continued to increase support for agricultural biotechnology research in the laboratory, even though only one transgenic crop, insect-resistant Bt cotton, is planted commercially on a large scale. In spite of progress over the last several decades, attempts to develop and deliver the products of modern agricultural biotechnology in China still remain fraught with challenges in the realms of research coordination, seed delivery, biosafety, and trade. The realization of sustained benefits from agricultural biotechnology in China will depend on whether or not these challenges can be overcome.

References

Athwal, D. S. (1971). Semi-dwarf rice and wheat in global food needs. *The Quarterly Review of Biology, 46*(1), 1–34.

Avery, O. T., MacLeod, C. M., & McCarty, M. (1944). Studies on the chemical nature of the substance inducing transformation of pneumococcal types: Induction of transformation by a desoxyribonucleic acid fraction isolated from Pneumococcus Type III. *Journal of Experimental Medicine, 79*(2), 137–158.

Beachy, R. N., Chen, Z.-L., Horsch, R. B., Rogers, S. G., Hoffman, N. J., & Fraley, R. T. (1985). Accumulation and assembly of soybean β-conglycinin in seeds of transformed petunia plants. *The EMBO Journal, 4*(12), 3047–3053.

Becker, J. (1998). *Hungry ghosts: Mao's secret famine.* New York, NY: Simon & Schuster.

Berg, P., Baltimore, D., Brenner, S., Roblin, R. O., & Singer, M. F. (1975). Summary statement of the Asilomar Conference on recombinant DNA molecules. *Proceedings of the National Academy of Sciences, 72*(6), 1981–1984.

Berg, P. & Singer, M. F. (1995). The recombinant DNA controversy: Twenty years later. *Proceedings of the National Academy of Sciences, 92*(20), 9011–9013.

Birchler, J. A., Auger, D. L., & Riddle, N. C. (2003). In search of the molecular basis of heterosis. *The Plant Cell, 15*(10), 2236–2239.

Bohannon, J. (2002). Zambia rejects GM corn on scientists' advice. *Science, 298*(5596), 1153–1154.

Bray, F. (2000). *Technology and society in Ming China (1368–1644).* Washington, D.C.: American Historical Association.

Brown, L. R. (1995). *Who will feed China? Wake-up call for a small planet.* New York, NY: W. W. Norton & Company.

Cao, C. (2004). Chinese science and the Nobel Prize complex. *Minerva, 42*(2), 151–172.

Carson, R. (1962). *Silent spring.* New York, NY: Houghton Mifflin Company.

Carter, C. A., Zhong, F., & Cai, F. (1996). *China's ongoing agricultural reform.* San Francisco, CA: 1990 Institute.

Chandler, R. F., Jr. (1982). *An adventure in applied science: A history of the International Rice Research Institute.* Los Baños, Laguna, Philippines: International Rice Research Institute.

Chang, K.-c. (1999). China on the eve of the historical period. In M. Loewe & E. L. Shaughnessy (Eds.), *Cambridge history of ancient China: From the origins of civilization to 221 B.C.* (pp. 37–71). Cambridge, United Kingdom: Cambridge University Press.

Chang, K.-c. (1986). *The archaeology of ancient China.* New Haven, CT: Yale University Press.

Charles, D. (2001). *Lords of the harvest: Biotech, big money, and the future of food.* Cambridge, MA: Perseus Publishing.

Chen, C., Sun, B., Pan, N., Liu, C., Liu, W., & Liang, X. (1992). An excellent virus-resistant transgenic fragnant tobacco [*sic*]. In Chen, Z. (Ed.), *Plant Genetic Engineering* (pp. 219–226). Beijing, China: Peking University Press.

Chen, H., Karplus, V. J., Ma, H., & Deng, X. W. (2006). Plant biology research comes of age in China. *The Plant Cell, 18*(11), 2855–2864.

Chen, S., Jin, W., Wang, M., Zhang, F., Zhou, J., & Jia, Q. et al. (2003). Distribution and characterization of over 1,000 T-DNA tags in rice genome. *The Plant Journal, 36*(1), 105–113.

Chen, S., Xu, C. G., Lin, X. H., & Zhang, Q. (2001). Improving bacterial blight resistance of '6078', an elite restorer line of hybrid rice, by molecular marker-assisted selection. *Plant Breeding, 120*(2), 133–137.

Chen, S. L., Cheng, B. L., Cheng, W. K., & Tang, P. S. (1945). An antibiotic substance in the Chinese water chestnut, *Eleocharis tuberose. Nature, 156,* 234.

Chen, Z. & Qu, L.-J. (1997). Plant molecular biology in China. *Plant Molecular Biology Reporter, 15*(3), 273–277.

China Daily. (1984, July 4). Group to study biotechnology. Beijing, China: Xinhua.

China National Center for Biotechnology Development (CNCBD). (2003). Biotechnology and bioindustry in China: History and vision of CNCBD. Beijing, China: CNCBD.

China Statistical Yearbook. (2006). Beijing, China: China Statistics Press.

CIA World Factbook. (2007). *China.* Washington, D.C.: U.S. Central Intelligence Agency. Retrieved April 24, 2007, from https://www.cia.gov/cia/publications/factbook/geos/ch.html.

Cleaver, H. M., Jr. (1972). The contradictions of the Green Revolution. *The American Economic Review, 62*(2), 177–186.

Cong, C. (2004). Chinese science and the Nobel Prize complex. *Minerva, 42*(1), 151–172.

Crook, F. (1988). Chapter 6. Agriculture. In R. L. Worden, A. M. Savada, & R. E. Dolan (Eds.), *China: A country study* [Electronic version]. Washington, D.C.: Library of Congress.

Crook, F. W. (1997). China: Is current ag policy a retreat from reform? *Agricultural Outlook Magazine* (Economic Research Service, USDA), 28–31. Retrieved April 24, 2007, from http://www.ers.usda.gov/publications/agoutlook/mar1997/ao238k.pdf.

de Brauw, A., Huang, J., & Rozelle, S. (2002). Sequencing and the success of gradualism: Empirical evidence from China's agricultural reform and the Chinese Academy of Sciences. Working Paper No. 02-005. Davis, CA: University of California at Davis, Department of Agricultural and Resource Economics.

Decree No. 40 State Council regulation on promoting the reform of the management system of seeds to strengthen market supervision. (2006). Bejing, China: State Council of the People's Republic of China.

Decree No. 62 Regulation on inspection and quarantine of import and export genetically modified commodities. (2001). Biosafety Clearing-House of China (National Biosafety Office, State Environmental Protection Administration, China). Retrieved April 24, 2007, from http://english.biosafety.gov.cn/image20010518/5516.pdf.

Decree No. 304 Regulations on safety of agricultural genetically modified organisms. (2001). Biosafety Clearing-House of China (National Biosafety Office, State Environmental Protection Administration, China). Retrieved April 24, 2007, from http://english.biosafety.gov.cn/image20010518/5420.pdf.

Ding, Y. (2006). Scientists' suicides prompt soul-searching in China. *Science, 311*(5763), 940–941.

Directive No. 349 of the Ministry of Agriculture of the People's Republic of China (in Chinese). (2004). Beijing, China: Ministry of Agriculture, People's Republic of China. Retrieved July 13, 2007, from http://www.agri.gov.cn/blgg/t20040223_168682.htm.

Directive 2001/18/EC of the European Parliament and of the Council of 12 March 2001 on the deliberate release into the environment of genetically modified organisms and repealing Council Directive 90/220/EEC. (2001). *Official Journal of the European Communities* [English electronic version], No L 106, 1–38.

Directive 90/219/EEC of the European Parliament and of the Council of 23 April 1990 on the contained use of genetically modified micro-organisms (1990). *Official Journal of the European Communities,* No L 117, 15–27.

Domes, J. (1980). *Socialism in the Chinese countryside: Rural societal policies in the People's Republic of China, 1949–1979.* London, United Kingdom: C. Hurst and Company.

Dong, D., Song, X., & Liu, R. (2007, January 29). Global Status of Biotech/GM Crops 2006 (Press Release - Chinese). Ithaca, NY: International Service for the Acquisition of Agri Biotech Applications. Retrieved August 22, 2007, from http://www.isaaa.org/resources/publications/briefs/35/pressrelease/pdf/Brief%2035%20-%20Press%20Release%20-%20Chinese.pdf.

Dunn, L. C. (1965). Mendel, his work and his place in history. *Proceedings of the American Philosophical Society, 109*(4), 189–198.

Economy, E. C. (2005). *The river runs black: The environmental challenge to China's future.* Ithaca, NY: Cornell University Press.

Elvin, M. (1973). *The pattern of the Chinese past.* Stanford, CA: Stanford University Press.

Evenson, R. E. (2004). Food and population: D. Gale Johnson and the Green Revolution. *Economic development and cultural change, 52*(3), 543–569.

Evenson, R. E. & Gollin, D. (2003). Assessing the impact of the Green Revolution, 1960–2000. *Science, 300*(5620) 758–762.

Ewen, S. W. & Pusztai, A. (1999). Effects of diets containing genetically modified potatoes expressing *Galanthus nivalis* lectin on rat small intestine. *The Lancet, 354*(9187), 1353–1354.

Falkner, R. (2006). International sources of environmental policy change in China: The case of genetically modified food. *The Pacific Review, 19*(4), 473–494.

Fan, S., Qian, K., & Zhang, X. (2006). China: An unfinished reform agenda. In P. G. Pardey, J. M. Alston & R. R. Piggott (Eds.), *Agricultural R & D in the developing world: Too little, too late?* (pp. 29–63). Washington, D.C.: International Food Policy Research Institute. Retrieved April 24, 2007, from http://www.ifpri.org/pubs/books/oc51/ oc51ch03.pdf.

Fan, S., Zhang, L., & Zhang, X. (2002). *Growth, inequality, and poverty in rural China: The role of public investments* (IFPRI Research Report No. 125). Washington, D.C.: International Food Policy Research Institute. Retrieved April 24, 2007, from http://www.ifpri.org/pubs/abstract/125/ab125.pdf.

Foreign Agricultural Service. (2006). People's Republic of China planting seeds annual. GAIN Report No. CH6104. Washington, D.C.: United States Department of Agriculture. Retrieved July 13, 2007, from http://www.fas.usda.gov/gainfiles/200612/146249710.pdf.

Foreign Agricultural Service. (2007). Cotton: World markets and trade (*Circular Series FOP 03–07*). Washington, D.C.: United States Department of Agriculture. Retrieved April 24, 2007, from http://www.fas.usda.gov/cotton/circular/2007/March/cottonfull0307.pdf.

Fu, C., Hu, W., Wang, Y., & Zhu, Z. (2005). Developments in transgenic fish in the People's Republic of China. *Scientific and Technical Review of the World Organization for Animal Health, 24*(1), 299–307. Retrieved July 13, 2007, from http://www.oie.int/eng/publicat/RT/2401/24-1%20pdfs/26-fu299-308.pdf.

Fuglie, K. O., Zhang, L., Salazar, L. F., & Walker, T. (1999). Economic impact of virus-free sweet potato planting material in Shandong province, China. (International Potato Center Publication Series). Lima, Peru: International Potato Center. Retrieved April 24, 2007, from http://www.eseap.cipotato.org/MF-ESEAP/Fl-Library/Eco-Imp-SP.pdf.

Giordano, M., Zhu, Z., Cai, X., Hong, S., Zhang, X., & Xue, Y. (2004). *Water management in the Yellow River Basin: Background, current critical issues and future research needs.* (Comprehensive Assessment Research Report Series No. 3). Colombo, Sri Lanka: Comprehensive Assessment Secretariat. Retrieved April 24, 2007, from http://www.iwmi.cgiar.org/assessment/FILES/pdf/publications/ResearchReports/CARR3.pdf.

Greenpeace International. (2005, April 13). Scandal: Greenpeace discovers illegal GE rice in China. *Greenpeace International News.* Retrieved April 24, 2007, from http://www.greenpeace.org/international/news/scandal-greenpeace-exposes-il.

Hamer, D. H. & Kung, S.-d. (1989). *Biotechnology in China.* Washington, D.C.: National Academies Press.

Hayami, Y. & Ruttan, V. W. (1985). *Agricultural development: An international perspective.* Baltimore, MD: Johns Hopkins University Press.

Hazell, P. B. R. (2003). The Green Revolution. In J. Mokyr (Ed.), *The Oxford encyclopedia of economic history.* Oxford, United Kingdom: Oxford University Press.

He, Z. H., Rajaram, S., Xin, Z. Y., & Huang, G. Z. (Eds.) (2001). *A History of Wheat Breeding in China*. Mexico: CIMMYT. Retrieved July 13, 2007, from http://www.cimmyt.org/english/docs/book/historywbchina/pdf/HistWBChina_contents.pdf.

Hellmich, R. L., Siegfried, B. D., Sears, M. K., Stanley-Horn, D. E., Daniels, M. J., & Mattila, H. R. et al. (2001). Monarch larvae sensitivity to *Bacillus thuringiensis*-purified proteins and pollen. *Proceedings of the National Academy of Sciences, 98*(21), 11925–11930.

Herman, E. (2003). Genetically modified soybeans and food allergies. *Journal of Experimental Botany, 54*(386), 1317–1319.

Ho, P.-T. (1956). Early ripening rice in Chinese history. *The Economic History Review, 9*(2), 200–218.

Ho, P.-T. (1975). *The cradle of the East: An inquiry into the indigenous origins of techniques and ideas of Neolithic and early historic China, 5000 – 1000 B.C.* Hong Kong: The Chinese University of Hong Kong.

Hossain, F., Pray, C., Liu, Y., Huang, J., Fan, C., & Hu, R. (2004). Genetically modified cotton and farmers' health in China. *International Journal of Occupational and Environmental Health, 10*(3), 296–303.

Hsu, C.-y. (1999). The Spring and Autumn Period. In M. Loewe & E. L. Shaughnessy (Eds.), *Cambridge history of ancient China: From the origins of civilization to 221 B.C.* (pp. 545–586). Cambridge, United Kingdom: Cambridge University Press.

Hsueh, Y. L. & Lou, C. H. (1947). Effect of 2,4-D on seed germination and respiration. *Science, 105*(2724), 283–285.

Hu, H., Dai, M., Yao, J., Xiao, B., Li, X., & Zhang, Q. et al. (2006). Overexpressing a NAM, ATAF, and CUC (NAC) transcription factor enhances drought resistance and salt tolerance in rice. *Proceedings of the National Academy of Sciences, 103*(35), 12987–12992.

Hu, R., Huang, J., Jin, S., & Rozelle, S. (2000). Assessing the contribution of research system and CG genetic materials to the total factor productivity of rice in China. *Journal of Rural Development, 23*(1), 33–70.

Hu, R., Shi, K., Cui, Y., & Huang, J. (2007). Change in China's agricultural research investment and its international comparison. *China Soft Science, 2*, 53–65.

Hu, W., Tong, S., Oldenburg, B., & Feng, X. (2001). Serum vitamin A concentrations and growth in children and adolescents in Gansu province, China. *Asia Pacific Journal of Clinical Nutrition, 10*(1), 63–66.

Huang, J. & Rozelle, S. (1996). Technological change: Rediscovering the engine of productivity growth in China's rural economy. *Journal of Development Economics, 49*(2), 337–369.

Huang, J. & Wang, Q. (2002). Agricultural biotechnology development and policy in China. *Agbioforum: The Journal of Agrobiotechnology Management and Economics, 5*(4), 122–135. Retrieved April 24, 2007, from http://www.agbioforum.org/v5n4/v5n4a01-huang.htm.

Huang, J., Hu, R., Fan, C., Pray, C., & Rozelle, S. (2002a). Bt cotton benefits, costs, and impacts in China. *Agbioforum: The Journal of Agrobiotechnology Management and Economics, 5*(4), 153–166. Retrieved April 24, 2007, from http://www.agbioforum.org/v5n4/v5n4a04-huang.htm.

Huang, J., Hu, R., Pray, C., & Rozelle, S. (2005a). *Development, policy and impacts of genetically modified crops in China: A comprehensive review of China's agricultural biotechnology sector.* (Presented at Workshop on Agricultural Biotechnology for Development: Institutional Challenges and Socioeconomic Issues in Bellagio, Italy, May 30 – June 1, 2005). Cambridge, MA: Science, Technology, and Public Policy Program.

Huang, J., Hu, R., Pray, C., Qiao, F., & Rozelle, S. (2003). Biotechnology as an alternative to chemical pesticides: A case study of Bt cotton in China. *Agricultural Economics, 29*(1), 55–67.

Huang, J., Hu, R., & Rozelle, S. (2002b). *Agricultural research investment in China: Challenges and prospects.* Beijing, China: Center for Chinese Agricultural Policy, Chinese Academy of Sciences.

Huang, J., Hu, R., Rozelle, S., & Pray, C. E. (2005b). Insect-resistant GM rice in farmers' fields: Assessing productivity and health effects in China. *Science, 308*(5722), 688–690.

Huang, J., Hu, R., Rozelle, S., Qiao, F., & Pray, C. E. (2002c). Transgenic varieties and productivity of smallholder cotton farmers in China. *The Australian Journal of Agricultural and Resource Economics, 46*(3), 1–21.

Huang, J., Qiu, H., Bai, J., & Pray, C. E. (2006). Awareness, acceptance and willingness to buy genetically modified foods in urban China. *Appetite, 46*(2), 144–151.

Huang, J., Rozelle, S., & Hu, R. (1998). Reforming China's seed industry: Transition to commercialization in the 21st century. *The Annual Report on Economic and Technological Development in Agriculture*. Beijing, China: Center for Chinese Agricultural Policy, Chinese Academy of Sciences.

Huang, J., Rozelle, S., Pray, C., & Wang Q. (2002d). Plant biotechnology in China. *Science, 295*(5555), 674–676.

Huang, J., Wang, Q., Zhang, Y., & Zepeda, J. F. (2001). *Agricultural biotechnology research indicators: China* (Supplementary material part d). Beijing, China: Center for Chinese Agricultural Policy: Chinese Academy of Sciences. Retrieved April 24, 2007, from http://www.sciencemag.org/cgi/data/295/5555/674/DC1/4.

Huang, Q., Rozelle, S., Howitt, R., Wang, J., & Huang, J. (2006). *Irrigation water pricing policy in China*. (Paper prepared for the American Agricultural Economics Association Annual Meeting in Long Beach, CA, July 23–26, 2006). Beijing, China: Center for Chinese Agricultural Policy, Chinese Academy of Sciences. Retrieved April 24, 2007, from http://agecon.lib.umn.edu/cgi-bin/pdf_view.pl?paperid=22362.

Huang, Y. (1998). *Agricultural reform in China: Getting institutions right*. Cambridge, United Kingdom: Cambridge University Press.

Hussain, A. (1989). Science and technology in the Chinese countryside. In D. F. Simon, & M. Goldman (Eds.), *Science and Technology in Post-Mao China* (pp. 223–249). Cambridge, MA: Harvard University Press.

International Rice Genome Sequencing Project. (2005). The Map-based sequence of the rice genome. *Nature, 436*(7052) 793–800.

James, C. (2006). Global status of commercialized biotech/GM crops: 2006 (ISAAA Brief No. 35, Executive Summary). Ithaca, NY: International Service for the Acquisition of Agri-Biotech Applications. Retrieved April 24, 2007, from http://www.isaaa.org/resources/publications/briefs/35/executivesummary/default.html.

Jia, H., Jayaraman, K. S., & Louet, S. (2004). China ramps up efforts to commercialize GM rice. *Nature Biotechnology, 22*(6), 642.

Jia, S. (2006). *Transgenic cotton*. Beijing, China: Science Press.

Jia, S.-R. & Jin, W.-J. (2002). The international debate on the biosafety of genetically modified crops: Scientific review of several cases of debate. *Chinese Journal of Agricultural Biotechnology, 1*(1), 3–8.

Jia, S. & Peng, Y. (2002). GMO biosafety research in China. *Environmental Biosafety Research, 1*(1), 5–8.

Jiang, G. H., Xu, C. G., Tu, J. M., Li, X. H., He, Y. Q., & Zhang, Q. F. (2004). Pyramiding of insect- and disease-resistance genes into an elite indica, cytoplasm male sterile restorer line of rice, 'Minghui 63'. *Plant Breeding, 123*(2), 112–116.

Jin, S., Rozelle, S., Alston, J. M., & Huang, J. (2005). Economies of scale and scope and the economic efficiency of China's agricultural research system. *International Economic Review, 46*(3), 1033–1057.

Jing, O. (2006, March 29). Scientific literacy: A new strategic priority for China. *Science and Development Network*. Retrieved July 13, 2007, from http://www.scidev.net/content/news/eng/scientific-literacy-a-new-strategic-priority-for-china.cfm.

Joravsky, D. (1961). The history of technology in Soviet Russia and Marxist doctrine. *Technology and Culture, 2*(1), 5–10.

Keeley, J. (2003). *The biotech developmental state? Investigating the Chinese gene revolution.* (Working Paper 207). Sussex, United Kingdom: Institute for Development Studies. Retrieved April 24, 2007, from http://www.ntd.co.uk/idsbookshop/details.asp?id=777.

Keeley, J. (2006). Balancing technological innovation and environmental regulation: An analysis of Chinese agricultural biotechnology governance. *Environmental Politics, 15*(2), 293–309.

Keightley, D. N. (1983). *The origins of Chinese civilization*. Berkeley, CA: University of California Press.

Krattiger, A. F. (1997). Insect resistance in crops: A case study of *Bacillus thuringiensis (Bt)* and its transfer to developing countries (ISAAA Brief No. 2). Ithaca, NY: International Service for the Acquisition of Agri-biotech Applications. Retrieved April 24, 2007, from http://www.isaaa.org/resources/publications/briefs/02/download/isaaa-brief-02-1997.pdf.

Kueh, Y. Y. (1993). Food consumption and peasant incomes. In Y. Y. Kueh & R. F. Ash (Eds.), *Economic trends in Chinese agriculture: The impact of post-Mao reforms* (pp. 229–272). Oxford, United Kingdom: Clarendon Press.

Kung, J. K. & Lin, J. Y. (2003). The causes of China's Great Leap famine, 1959–1961. *Economic development and cultural change, 52*(1) 51–73.

Lei, Y. (2006, March 5). Facts and figures: China's drive to build a new socialist countryside. *Gov.cn* (The Chinese government's official web portal). Retrieved April 24, 2007, from http://english.gov.cn/2006-03/05/content_218920.htm.

Lewis, M. E. (1999). Warring states: Political history. In M. Loewe & E. L. Shaughnessy (Eds.), *Cambridge history of ancient China: From the origins of civilization to 221 B.C.* (pp. 587–645). Cambridge, United Kingdom: Cambridge University Press.

Li, H.-L. (1983). The domestication of plants in China: Ecogeographical considerations. In D. Knightley (Ed.), *The origins of Chinese civilization* (pp. 21–64). Berkeley, CA: University of California Press.

Li, J., Zhu, S., Song, X., Shen, Y., Chen, H., & Yu, J. (2006). A rice glutamate receptor-like gene is critical for the division and survival of individual cells in the root apical meristem. *The Plant Cell, 18*(2), 340–349.

Li, L. M. (1982). Introduction: Food, famine, and the Chinese State. *The Journal of Asian Studies, 41*(4), 687–707.

Li, O., Li, X., Liu, Y., Wang, D., & Jian, X. (1995). Emerging roles of the universities and colleges in agricultural extension and rural development in China. *Journal of International Agricultural and Extension Education, 2*(2), 68–78. Retrieved July 13, 2007, from http://www.aiaee.org/archive/Vol-2.2.pdf.

Li, Q., Curtis, K. R., McCluskey, J. J., & Wahl, T. I. (2002). Consumer attitudes toward genetically modified foods in Beijing, China. *Agbioforum: The Journal of Agrobiotechnology Management and Economics, 5*(4), 145–152. Retrieved April 24, 2006, from http://www.agbioforum.org/v5n4/v5n4a03-wahl.htm.

Lin, J. Y. (1988). The Household Responsibility System in China's agricultural reform: A theoretical and empirical study. *Economic Development and Cultural Change* (Supplement: Why does Overcrowded, Resource-Poor East Asia Succeed: Lessons for the LDCs?), *36*(3), S199–S224.

Lin, J. Y. (1990). Collectivization and China's agricultural crisis in 1959-1961. *The Journal of Political Economy, 98*(6), 1228–1252.

Lin, J. Y. (1992). Rural reforms and agricultural growth in China. *American Economic Review, 82*(1), 34–51.

Lin, W., Tuan, F., Dai, Y., & Zhong, F. (2007). Does biotech labeling affect consumers' purchasing decisions? A case study of vegetable oils in Nanjing, China. (Prepared for the International Agricultural Trade and Research Consortium summer meeting in Beijing, China, July 7–10, 2007). Washington, D.C.: United States Department of Agriculture Economic Research Service.

Lin, Z., Hu, Y., & Ni, T. (2004). Overview of research on transgenic potato and their biosafety in China. (Proceedings of the Sixth World Potato Congress, Boise, ID, August 20–26, 2006). Retrieved April 24, 2006, from http://www.potatocongress.org/sub.cfm?source=289.

Liu, Y.-B., Tabashnik, B., Dennehy, T., Patin, A., & Bartlett, A. (1999). Development time and resistance to Bt crops. *Nature, 400*(6744), 519.

Loewe, M. & Shaughnessy, E. L. (Eds.) (1986). *Cambridge history of ancient China: From the origins of civilization to 221 B.C.* Cambridge, United Kingdom: Cambridge University Press.

Losey, J. E., Rayor, L. S., & Carter, M. E. (1999). Transgenic pollen harms monarch larvae. *Nature, 399*(6733), 214.

Lou, C. H. (1945). Fluorescein-induced parthenocarpy. *Nature, 155*, 23.

Macilwain, C. (2003). Chinese agribiotech: Against the grain. *Nature, 422*(6928), 111–112.

Malakata, M. (2007, April 12). Zambia takes steps towards biosafety law. *Science and Development Network*. Retrieved April 24, 2007, from http://www.scidev.net/dossiers/index.cfm?fuseaction=dossierreaditem&dossier=6&type=1&itemid=3549&language=1.

Matson, P. A., Parton, W. J., Power, A. G., & Swift, M. J. (1997). Agricultural intensification and ecosystem properties. *Science, 277*(5325), 504–509.

Mendel, G. (1866). Versuche über Pflanzenhybriden. *Verhandlungen des narturforschenden Vereines in Brünn, Bd. IV für das Jahr 1865* (Electronic Scholarly Publishing Project, 1996). Retrieved April 24, 2007, from http://www.esp.org/foundations/genetics/classical/gm-65.pdf.

Ministry of Agriculture (P. R. China) Directive No. 349. (2004). Ministry of Agriculture, People's Republic of China. Retrieved April 24, 2007, from http://www.agri.gov.cn/blgg/t20040223_168682.htm.

Morris, C. E. & Sands, D. C. (2006). The breeder's dilemma – yield or nutrition? *Nature Biotechnology, 24*(9), 1078–1080.

Morris, M. L. (2002). *Impacts of international maize breeding research in developing countries, 1966–98*. Mexico: CIMMYT. Retrieved April 24, 2007, from http://www.cimmyt.org/Research/Economics/map/impact_studies/maize1966_98/m66_98contents.pdf.

Myers, R. H. (1970). *The Chinese peasant economy: Agricultural development in Hopei and Shantung, 1890–1949*. Cambridge, MA: Harvard University Press.

Nao, N. (2005, March 11). China close to production of "safe" genetic rice. [Electronic version]. *Reuters*. Retrieved April 24, 2007, from Factiva online database.

Needham, J. (1984). *Science and civilization in China*. Cambridge, United Kingdom: Cambridge University Press.

Nordlee, J. A., Taylor, S. L., Townsend, J. A., Thomas, L. A., & Bush, R. K. (1996). Identification of a Brazil nut allergen in transgenic soybeans. *The New England Journal of Medicine, 334*(11), 688–692.

Normile, D. (1999). Rice biotechnology: Rockefeller to end network after 15 years of success. *Science, 286*(5444), 1468–1469.

Oberhauser, K. S., Prysby, M. D., Mattila, H. R., Stanley-Horn, D. E., Sears, M. K., & Dively, G. et al. (2001). Temporal and spatial overlap between monarch larvae and corn pollen. *Proceedings of the National Academy of Sciences, 98*(21), 11913–11918.

Oi, J. C. (1999). *Rural China takes off: Institutional foundations of economic reform*. Berkeley, CA: University of California Press.

Oldham, G. E. (1997). *A decade of reform: Science and technology policy in China*. Ottawa, Canada: International Development Research Center. Retrieved April 24, 2007, from http://www.idrc.ca/en/ev-9360-201-1-DO_TOPIC.html.

O'Toole, J. C., Toenniessen, G. H., Murashige, T., Harris, R. R., & Herdt, R. W. (2000). The Rockefeller Foundation's International Program on Rice Biotechnology. In G. S. Khush, D. S. Brar, & B. Hardy (Eds.), *Proceedings of the fourth international rice genetics symposium* (pp. 39–60). Los Baños, Laguna, Philippines: International Rice Research Institute. Retrieved April 24, 2007, from http://www.rockfound.org/library/01rice_bio.pdf.

People's Daily Online. (2006, April 22). China embraces new scientific development concept: Hu. Beijing, P.R. China: Xinhua. Retrieved April 24, 2007, from http://english.peopledaily.com.cn/200604/22/eng20060422_260256.html.

Piazza, A. & Liang, E. H. (1998). Reducing absolute poverty in China: Current status and issues. *Journal of International Affairs, 52*(1), 253–274.

Pimentel, D., Berger, B., Filiberto, D., Newton, M., Wolfe, B., & Karabinakis, E. et al. (2004). Water resources: Agricultural and environmental issues. *Bioscience, 54*(10), 909–918.

Pinstrup-Andersen, P. & Schioler, E. (2001). *Seeds of contention: World hunger and the global controversy over genetically modified foods*. Washington, D.C.: International Food Policy Research Institute.

Pleasants, J. M., Hellmich, R. L., Dively, G. P., Sears, M. K., Stanley-Horn, D. E., & Mattila, H. R. et al. (2001). Corn pollen deposition on milkweeds in and near cornfields. *Proceedings of the National Academy of Sciences, 98*(21), 11919–11924.

Pray, C. E. (1999). Public and private collaboration on plant biotechnology in China. *Agbioforum: The Journal of Agrobiotechnology Management and Economics, 2*(1), 48–53. Retrieved April 24, 2007, from http://www.agbioforum.org/v2n1/v2n1a09-pray.pdf.

Pray, C. E. (2001a). China. In C. E. Pray & K. O. Fuglie (Eds.), *Private investment in agricultural research and international technology transfer in Asia* (pp. 137–155) (Agricultural Economics Report No. AER805). Washington, D. C.: United States Department of Agriculture.

Pray, C. E., Huang, J., Hu, R., & Rozelle, S. (2002). Five years of Bt cotton in China: The benefits continue. *The Plant Journal, 31*(4), 423–430.

Pray, C., Ma, D., Huang, J., & Qiao, F. (2001b). Impact of Bt cotton in China. *World Development, 29*(5), 813–825.

Pray, C. E., Ramaswami, B., Huang, J., Hu, R., Bengali, P., & Zhang, H. (2006). Costs and enforcement of biosafety regulations in India and China. *International Journal of Technology and Globalisation, 2*(1/2), 137–157.

Qiao, F., Lohmar, B., Huang, J., Rozelle, S., & Zhang, L. (2003). Producer benefits from input market and trade liberalization: The case of fertilizer in China. *American Journal of Agricultural Economics, 85*(5), 1223–1227.

Regulation No. 1829/2003 of the European Parliament and of the Council of 23 April 1990 on the contained use of genetically modified micro-organisms [English electronic version]. (2001). *Official Journal of the European Communities,* L 268–46, 1–23.

Regulation No. 1830/2003 of the European Parliament and of the Council of 22 September 2003 concerning the traceability and labeling of genetically modified organisms and the traceability of food and feed products produced from genetically modified organisms and amending Directive 2001/18/EC [English electronic version]. (2001). *Official Journal of the European Communities,* L 268–46, 24–28.

Riley, R. & Constabel, F. (2006). *Breeding (plant)* [Electronic Version]. McGraw-Hill Encyclopedia of Science and Technology Online.

Rockefeller Foundation. (2006). *Africa's turn: A new Green Revolution for the 21st century.* Retrieved April 24, 2007, from http://www.rockfound.org/library/africas_turn.pdf.

Ruttan, V. W. (2002). Productivity growth in world agriculture: Sources and constraints. *Journal of Economic Perspectives, 16*(4), 161–184.

Scott, D. (2003). Science and the consequences of mistrust: Lessons from recent GM controversies. *Journal of Agricultural and Environmental Ethics, 16,* 569–582.

Sears, M. K., Hellmich, R. L., Stanley-Horn, D. E., Oberhauser, K. S., Pleasants, J. M., & Mattila, H. R. et al. (2001). Impact of Bt corn pollen on Monarch butterfly populations: A risk assessment. *Proceedings of the National Academy of Sciences, 98*(21), 11937–11942.

Seed Law of the Peoples Republic of China. (2000). Beijing, China: Legislative Affairs Commission of the Standing Committee of the Ninth National People's Congress of the People's Republic of China. Retrieved April 24, 2007, from http://english.gov.cn/laws/2005-09/08/content_30273.htm.

Segal, A. (2003). *Digital dragon: High-technology enterprises in China.* Ithaca, NY: Cornell University Press.

Sinha, G. (2007). Agbiotech: GM technology develops in the developing world. *Science, 315*(5809), 182–183.

Smil, V. (1997). Global population and the nitrogen cycle. *Scientific American, 277*(1), 76–82.

Smil, V. (2004). *China's past, China's future: Energy, food, environment.* New York, NY: RoutledgeCurzon.

Song, R. (2001). *Du Jiang Yan irrigation system.* Chengdu, China: Tian Di Publishing House.

Song, W. Y., Wang, G. L., Chen, L., Kim, K. S., Holsten, T., & Gardner, J. et al. (1995). A receptor kinase-like protein encoded by the rice disease resistance gene, *Xa21. Science, 270*(5243): 1804–1806.

Song, Y. (1999). Introduction of transgenic cotton in China. *Biotechnology and Development Monitor, 37,* 14–17.

Stanley-Horn, D. E., Dively, G. P., Hellmich, R. L., Mattila, H. R., Sears, M. K., & Rose, R. et al. (2001). Assessing the impact of Cry1 Ab-expressing corn pollen on monarch butterfly larvae in field studies. *Proceedings of the National Academy of Sciences, 98*(21), 11931–11936.

Stone, B. (1990). Evolution and diffusion of agricultural technology in China. In N. G. Kotler (Ed.), *Sharing innovation: Global perspectives on food, agriculture, and rural development* (pp. 35–93). Washington, D. C.: Smithsonian Institution Press.

Stryer, L. (1995). *Biochemistry (4th ed.)*. New York, NY: W. H. Freeman and Company.

Su, J., Xiao, Y., Li, M., Liu, Q., & Li, B. et al. (2006). Mapping QTLs for phosphorus-deficiency tolerance at wheat seedling stage. *Plant and Soil, 281*(1–2), 25–36.

Suttmeier, R. P. & Cao, C. (1999). China faces the new industrial revolution: Achievement and uncertainty in the search for innovation strategies. *Asian Perspective, 23*(3), 153–200.

Suttmeier, R. P., Cao, C., & Simon D. F. (2006). "Knowledge innovation" and the Chinese Academy of Sciences. *Science, 312*(5770), 58–59.

Tang, P.-S. (1981). Regulation and control of multiple pathways of respiratory metabolism in relation to other physiological functions in higher plants: Recollections and reflections on fifty years of research in plant respiration. *American Journal of Botany, 68*(3), 443–448.

Tang, P.-S. (1983). Aspirations, reality, and circumstances: The devious trail of a roaming plant physiologist. *Annual Review Plant Physiology, 34*, 1–20.

Tang, P.-S. & Loo, S.W. (1940). Polyploidy in soybean, pea, wheat and rice, induced by colchicines treatment. *Science, 91*(2357), 222.

Tang, P.-S. & Wu, H.-Y. (1957). Adaptive formation of nitrate reductase in rice seedlings. *Nature, 179*(4574), 1355–1356.

Tang, T. B. (1984). *Science and Technology in China*. London, United Kingdom: Longman Group Limited.

Tawney, R. H. (1964). *Land and Labour in China*. New York, NY: Octagon Books.

Teng, K. (2004). The commercial environment and product development for agricultural biotechnology in China. Working Paper No. 04–E3. Beijing, China: Center for Chinese Agricultural Policy. Retrieved August 22, 2007, from http://www.ccap.org.cn/PDF/WP-04-E3.pdf.

Thompson, J. A. (2002). *Genes for Africa: Genetically Modified Crops in the Developing World*. Cape Town, South Africa: University of Cape Town Press.

Tilman, D. (1999). Global Environmental Impacts of Agricultural Expansion: The Need for Sustainable and Efficient Practices. *Proceedings of the National Academy of Sciences, 96*(11) 5995–6000.

Tu, J., Ona, I., Zhang, Q., Mew, T. W., Khush, G. S. & Datta, S. K. (1998). Transgenic rice variety 'IR72' with Xa21 is resistant to bacterial blight. *Theoretical and Applied Genetics, 97*(1–2), 31–36.

USDA Office of the Chief Economist. (2006). Cotton: China. *Weather and climate: Major world crop areas and climatic profiles*. Washington, D.C.: United States Department of Agriculture. Retrieved April 24, 2007, from http://www.usda.gov/oce/weather/pubs/Other/MWCACP/Graphs/chi/chicot.gif.

Virmani, S. S., Sun, Z. X., Mou, T. M., Ali, A. J., & Mao, C. X. (2003). *Two-line hybrid rice breeding manual*. Los Baños, Laguna, Philippines: International Rice Research Institute. Retrieved April 24, 2007, from http://www.knowledgebank.irri.org/hybridRiceSeed/twoLineHybridRiceBreedingManual.pdf.

Wang, G.-L., Ruan, D.-L., Song, W.-Y., Sideris, S., Chen, L., & Pi, L.-Y. et al. (1998). *Xa21D* encodes a receptor-like molecule with a leucine-rich repeat domain that determines race-specific recognition and is subject to adaptive evolution. *The Plant Cell, 10*(5), 765–779.

Wang, S., Liu, N., Peng, K., & Zhang, Q. (1999). The distribution and copy number of *copia*-like retrotransposons in rice (*Oryza sativa* L.) and their implications in the organization and evolution of the rice genome. *Proceedings of the National Academy of Sciences, 96*(12), 6824–6828.

Wang, X. & Liu, Y. (2005). Technological progress and Chinese agricultural growth in the 1990s. *China Economic Review, 16*(4), 419–440.

Wang, Z.-Y., Zheng, F.-Q., Shen, G.-Z., Gao, J.-P., Snustad, D. P., & Li, M.-G., et al. (1995). The amylose content in rice endosperm is related to the post-transcriptional regulation of the *waxy* gene. *The Plant Journal, 7*(4), 613–622.

Wilsdon, J. & Keeley, J. (2007). *China: The next science superpower?* London, United Kingdom: Demos.

WTO (World Trade Organization). (1994). Sanitary and phytosanitary measures: Text of the agreement (Article 3.3). Retrieved April 24, 2007, from http://www.wto.org/english/tratop_e/sps_e/spsagr_e.htm.

Wong, R. B. (1982). Food riots in the Qing Dynasty. *The Journal of Asian Studies, 41*(4), 767–788.

Wu, F. & Butz, W. P. (2004). *The future of genetically modified crops: Lessons from the Green Revolution.* Santa Monica, CA: RAND Corporation.

Wu, K. (2002). *A brief statement on the studies of the ecological impact of Bt cotton conducted by Dr. Kongming Wu's lab* (Unpublished manuscript). Beijing, China: Institute of Plant Protection, Chinese Academy of Agricultural Sciences.

Wu, K. M. & Guo, Y. Y. (2005). The evolution of cotton pest management practices in China. *Annual Review of Entomology, 50,* 31–52.

Wu, K., Guo, Y., & Lv, N. (1999). Geographic variation in susceptibility of *Helicoverpa armigera* (Lepidoptera: Noctuidae) to *Bacillus thuringiensis* insecticidal protein in China. *Journal of Economic Entomology, 92*(2), 273–278.

Wu, K., Guo, Y., Lv, N., Greenplate, J. T., & Deaton, R. (2003). Efficacy of transgenic cotton containing a *cry1Ac* gene from *Bacillus thuringiensis* against *Helicoverpa armigera* (Lepidoptera: Noctuidae) in northern China. *Journal of Economic Entomology, 96*(4), 1322–1328.

Wu, K., Lin, K., Miao, J., & Zhang, Y. (2005). Field abundances of insect predators and insect pests on d-endotoxin-producing transgenic cotton in northern China. In M. S. Hoddle (ed.), *Proceedings of the Second International Symposium on the Biological Control of Arthropods, Vol. 1* (pp. 362–368), Washington, D.C.: U. S. Department of Agriculture, Forest Service.

Wu, R. (1983, January 10). Letter from Dr. Ray Wu to Chinese State Councillor Fang Yi. Copy of original provided by author.

Xiong, V. C. (2006). *Emperor Yang of the Sui Dynasty: His life, times, and legacy.* Albany, NY: State University of New York Press.

Xue, D. (2002, June 4). *A summary of research on the environmental impact of Bt Cotton in China* (Unpublished manuscript). Greenpeace International.

Yabuuti, K. (1967). *T'ien-kung K'ai-wu: Chinese Technology in the Seventeenth Century* (Book Reviews). *Technology and Culture, 8*(1), 92–94.

Yan, X., Wu, P., Ling, H., Xu, G., Xu, F., & Zhang, Q. (2006). Plant nutriomics in China: An overview. *Annals of Botany, 98*(3), 473–482.

Yang, L., Jiang, J., Wei, W., Zhang, B., Wang, L., & Yang, S. (2006). The *pha2* gene cluster involved in Na+ resistance and adaption to alkaline pH in *Sinorhizobium fredii* RT19 encodes a monovalent cation/proton antiporter. *FEMS Microbiology Letters, 262*(2), 172–177.

Yin, H. C., & Sun, C. N. (1947). Histochemical method for the detection of phosphorylase in plant tissues. *Science, 105*(2738), 650.

Yin, H. C. & Tung, Y. T. (1948). Phosphorylase in guard cells. *Science, 108*(2795), 87–88.

Yue, B., Xue, W., Xiong, L., Yu, X., Luo, L., & Cui, K. et al. (2006). Genetic basis of drought resistance at reproductive stage in rice: Separation of drought tolerance from drought avoidance. *Genetics, 172*(2) 1213–1228.

Zangerl, A. R., McKenna, D., Wraight, C. L., Carroll, M., Ficarello, P., & Warner, R., et al. (2001). Effects of exposure to event 176 *Bacillus thuringiensis* corn pollen on monarch and black swallowtail caterpillars under field conditions. *Proceedings of the National Academy of Sciences, 98*(21), 11908–11912.

Zhang, J., Guo, D., Chang, Y., You, C., & Li, X. et al. (2007). Non-random distribution of T-DNA insertions at various levels of the genome hierarchy as revealed by analyzing 13 804 T-DNA flanking sequences from an enhancer-trap mutant library. *The Plant Journal, 49*(5), 947–959.

Zhang, L., de Brauw, A., & Rozelle, S. (2004). China's rural labor market development and its gender implications. *China Economic Review, 15*(2), 230–247.

Zhang, Q. & Luo, L. (1999). Improving the tolerance of irrigated rice to water-stressed conditions. In J.-M. Ribaut & D. Poland (Eds.), *Molecular approaches for the genetic improvement of cereals for stable production in water-limited environments* (Workshop, El Batan, Mexico, June 21-25, 1999). Mexico: CIMMYT. Retrieved April 24, 2007, from http://www.cimmyt.org/english/docs/proceedings/molecApproaches/pdfs/improving_tolerance.pdf.

Zhang, Q., Shen, B. Z., Dai, X. K., Mei, M. H., Saghai Maroof, M. A., & Li, Z. B. (1994). Using bulked extremes and recessive class to map genes for photoperiod-sensitive genic male sterility in rice. *Proceedings of the National Academy of Sciences, 91*(18), 8675—8679.

Zhang, Y., Pan, D., Sun, X., Sun, G., Wang, X., & Liu, X., et al. (2006). Production of porcine cloned transgenic embryos expressing green fluorescent protein by somatic cell nuclear transfer. *Science in China Series C: Life Sciences, 49*(1), 1–8.

Zhao, R. (2003). Transition in R&D management control system: Case study of a biotechnology research institute in China. *Journal of High Technology Management Research, 14*(2), 213–229.

Zhong, F., Marchant, M. A., Ding, Y., & Lu, K. (2002). GM foods: A Nanjing case study of Chinese consumers' awareness and potential attitudes. *Agbioforum: The Journal of Agrobiotechnology Management and Economics, 5*(4), 136–144: Retrieved July 11, 2007, from http://www.agbioforum.org/v5n4/v5n4a02-zhong.pdf.

Zhong, T., Rozelle, S., Stone, B., Jiang D., Chen, J., & Xu, Z. (1995). China's experience with market reform for commercialization of agriculture in poor areas. In J. Von Braun & E. Kennedy (Eds.), *Agricultural commercialization, economic development, and nutrition* (pp. 119–140). Washington D.C.: International Food Policy Research Institute.

Zhu, Z. L. & Chen, D. L. (2002). Nitrogen fertilizer use in China – Contributions to food production, impacts on the environment and best management strategies. *Nutrient Cycling in Agroecosystems, 63*(2/3), 117–127.

Zhang, M. Binns, C. W., & Lee, A. H. (2002). Dietary patterns and nutrient intake of adult women in south-east China: A nutrition study in Zhejiang province. *Asia Pacific Journal of Clinical Nutrition, 11*(1), 13–21.

Zi, X. (2005). GM rice forges ahead in China amid concerns over illegal planting. *Nature Biotechnology, 23*(6), 637.

About the Authors

Valerie J. Karplus graduated from Yale University in 2002 with a Bachelor of Science degree in Molecular Biophysics and Biochemistry and Political Science. She lived in China for two years, where she researched the development and impact of agricultural biotechnology in China while based at the China Agricultural University (2002–2003) and the National Institute of Biological Sciences, Beijing (2005–2006). Currently she is pursuing graduate study at the Massachusetts Institute of Technology.

Dr. Xing Wang Deng is the Daniel C. Eaton Professor of Plant Biology at Yale University. His scientific work focuses on the molecular and genomic basis for plant development and agricultural biotechnology. He also serves as a the co-director of the National Institute of Biological Sciences, Beijing and the founding director of the Peking-Yale Joint Research Center of Plant Molecular Genetics and Agro-biotechnology. He leads a research team that has published over a hundred peer-reviewed articles in his area of research.

Index

Printed in the United States
By Bookmasters